Science Education for Everyday Life

Evidence-Based Practice

Glen S. Aikenhead

TEACHERS COLLEGE PRESS

Teachers College
Columbia University
New York and London

KH

Published by Teachers College Press, 1234 Amsterdam Avenue, New York, NY 10027

Chapter 3 is based on a much shorter piece, "Humanistic Perspectives in the Science Curriculum," which appears as a chapter in the *Handbook of Research on Science Education,* edited by S. K. Abell and N. G. Lederman, to be published by Lawrence Erlbaum Associates in 2006.

Acknowledgement: The author wishes to acknowledge the University of Saskatchewan's Publishing Subvention grant which supported in part the publication of this book.

Library of Congress Cataloging-in-Publication Data

Aikenhead, Glen S.
    Science education for everyday life : evidence-based practice / Glen S. Aikenhead.
        p. cm. — (Ways of knowing in science and mathematics series)
    Includes bibliographical references and index.
    ISBN 0-8077-4634-7 (alk. paper)
    1. Science—Study and teaching.   I. Title.   II. Series.

    Q181.A36 2005
    507'.1'273—dc22

                                                                    2005050627

ISBN-13: 978-0-8077-4634-9 (paper)   ISBN-10: 0-8077-4634-7 (paper)

Printed on acid-free paper

Manufactured in the United States of America

13   12   11   10   09   08   07   06   8   7   6   5   4   3   2   1

6/23/06

Contents

1

Introduction

By convention, school science has traditionally attempted to prepare students for the next level of science courses by focusing on intellectual knowledge acquisition. Its ultimate purpose has been to funnel capable students into science and engineering degree programs, a phenomenon or ideology often called "the pipeline" (DeBoer, 1991; Frederick, 1991; Millar & Osborne, 1998). The traditional science curriculum advocates canonical science content and habits of mind (i.e., thinking and believing like a scientist). This scientist-oriented approach assumes that "science" in "school science" has the same narrow meaning as it has in, for example, "the American Association for the Advancement of Science." A different assumption for school science is explored in this book.

Most high school students embrace career goals outside the pipeline and often do not feel comfortable in its scientist-oriented ideology (Atkin & Helms, 1993; Eisenhart, Finkel, & Marion, 1996; Kozoll & Osborne, 2004; Lyons, 2003; McKinley, McPherson, & Bell, 1992; Reiss, 2000). These students experience school science as a foreign culture (Calabrese Barton, 2002; Costa, 1995; Thijs & van den Berg, 1995). They would prefer a science education for everyday life. Educators inevitably rationalize pipeline school science as serving two main purposes for these students: their need to understand science well enough to appreciate its national importance, and their need to be literate enough to receive scientific messages expressed by scientific experts.

Alternative rationales for the science curriculum have existed since its inception. The most pervasive alternative has been the perspective that celebrates science as a human endeavor, embedded within a social milieu of society and conducted by various social communities of scientists. This view of science is endemic to a long-standing purpose for school science: to develop students' capacities to function as responsible savvy participants in their everyday lives increasingly affected by science and technology. To function responsibly, people must be able to treat science and technology as "a repository to be raided for what it can contribute to the achievement

1

of practical ends" (Layton, Jenkins, Macgill, & Davey, 1993, p. 135) such as science-related daily activities, personal problems, social issues, or global concerns. To engage with science and technology toward practical ends, people must be able to critically assess the information they come across and critically evaluate the trustworthiness of the information source. A critical appreciation of science is central to this alternative perspective, for all students.

Inspired by ideologies broadly associated with educationalists such as John Dewey (1916) and Harold Benjamin (1939), a minority of science educators has always embraced this alternative everyday-life approach that animates students' self-identities, their future contributions to society as citizens, and their interest in making personal utilitarian meaning of scientific and technological knowledge. These science educators have written thoughtful apologias for *what ought to happen* in science classrooms if such a student-oriented ideology is to be realized. This book follows a different agenda. It focuses on *the results of research* into student-oriented school science that treats science as a human endeavor. This departure from past agendas is significant in that this book will generate evidence-based findings rather than philosophical rationales.

This alternative to the traditional science curriculum was called "humanistic" by Duschl (1988) and Solomon (1997), among others, on the basis of their analysis of the curriculum literature. Donnelly's (2004) recent historical account of the term *humanistic* in the context of science curricula illustrates its diverse and sometimes contested application to curriculum development. Although the word *humanistic* can have different meanings in different countries and in various scholarly disciplines, and although it has several connotations within education itself, I find *humanistic* the best word to describe the diverse yet pervasive alternative to the pipeline ideology of traditional school science.

Presignifying Chapter 2, let me clarify this book's broad meaning of *humanistic*. In the science education research literature, humanistic perspectives have referred to values, the nature of science, the social aspects of science, the culture of science, and the human character of science revealed through its sociology, history, and philosophy. Since the 1970s, humanistic perspectives were typically found in science-technology-society (STS) curricula, a field reviewed by Solomon and me in 1994. But humanistic perspectives are not restricted to STS curricula. Table 1.1 describes dichotomies of goals, values, and ideologies found in various research studies published in the science education literature. Each row in Table 1.1 represents what is generally included and excluded by the phrase "humanistic perspectives" as used in this book. The various rows indicate different definitions found in the research literature. The first column in Table 1.1

TABLE 1.1. Possible Characteristics of a Humanistic Perspective in School Science

Included	Excluded
Induction, socialization, or enculturation into students' local, national, and global communities that are increasingly shaped by science and technology.	Induction, socialization, or enculturation into a scientific discipline.
Citizenship preparation for the everyday world.	Preprofessional training for the scientific world.
Savvy citizens cognizant of the human, social, and cultural dimensions of scientific practice and its consequences.	Canonical abstract ideas (curricular content) most often decontextualized from everyday life but sometimes placed in a trivial everyday context.
Attention to several sciences: established science, frontier science, and/or citizen science.	Emphasis on established science only.
Multiscience approach reflecting international perspectives (including indigenous science).	Mono-science approach founded on universalism (Western science).
Knowledge *about* science and scientists.	Knowledge of canonical science.
Moral reasoning integrated with values, human concerns, and scientific reasoning.	Solely scientific reasoning using scientific habits of mind.
Seeing the world through the eyes of students and significant adults.	Seeing the world through the eyes of scientists alone.
Learning is interacting with the everyday world and includes intellectual achievement, personal change, forming new self-identities, recognizing sociopolitical power, and perhaps practical or social action.	Learning is an intellectual task focused on acquiring scientific knowledge and scientific habits of mind.
Playing in the subculture of science as an outsider.	Identifying with the subculture of science as an insider.

represents possible humanistic perspectives in a humanistic science course, whereas the second column describes a traditional science course. Humanistic science courses have usually combined some of column 1 with some of column 2, in order to meet the needs of students and the expectations of schools, parents, universities, and other stakeholders. This integration of humanistic and canonical science perspectives varies with different research studies (Aikenhead, 1994d; Bartholomew, Osborne, & Ratcliffe, 2004). Terminology in Table 1.1, such as "citizen science" and "indigenous science," will be clarified in the context of the research synthesized in subsequent chapters.

In this book I synthesize research that has explored humanistic perspectives in school science, perspectives that significantly alter the tenor of traditional school science. When evidence-based findings are synthesized, trends in the successes and failures of humanistic approaches become apparent. These trends can inform the decisions made at the policy level, the classroom level, and the teacher education level, all focused on evidence-based practice. Strengths and weaknesses in the research itself will be identified, thereby providing the reader with fruitful ideas for further research.

Any perspective on school science, be it humanistic or solely scientific, conveys an ideological point of view (Cross, 1997a; Fensham, 2000b; Fourez, 1989; Knain, 2001; Säther, 2003). This book's ideology gives priority to a student-oriented point of view aimed at citizens acting as consumers of science and technology in their everyday lives. A different ideology, mentioned above, is an elite scientist-oriented view aimed at marshaling "the best and brightest" through the pipeline, an ideology of preprofessional training.

In the political arena defined by Spencer's (1859) key question, "What knowledge is of most worth?" (p. 5), the research literature expresses essentially two contrary positions, often in combination (Lijnse, Kortland, Eijkelhof, van Genderen, & Hooymayers, 1990; Solomon, 1999b). On the one hand, the research focuses on educationally driven propositions about what is best for students and society; but on the other hand, the research meets politically driven realities that counteract research findings. For instance, empirical evidence overwhelmingly speaks to the *educational* failure of traditional school science assessed by its own criteria (Chapter 3), but the continuous survival and high status enjoyed by traditional school science attest to its *political* success.

The research discussed in this book reflects the tension between educational soundness and political reality. We must never forget that curriculum decisions (What knowledge is of most worth? and Who decides?) are first and foremost political decisions (Atwater, 1996; Calabrese Barton

& Osborne, 2001; Eijkelhof & Kapteijn, 2000; Fensham, 1992; Gaskell, 2003; Östman, 1996; Roberts, 1988; Rudolph, 2003). Although research has a role to *inform* curriculum decisions concerning a humanistic science curriculum, supportive evidence-based findings from research tend to wilt in the glare of opposing ideologies. Curriculum choices in the past have most often favored the status quo over a proposed humanistic science curriculum (Carlone, 2003; Fensham, 1992, 1993, 1998; Gaskell, 1992, 2003; Hart, 2002; Hurd, 1991; Klopfer, 1992; Panwar & Hoddinott, 1995; Walberg, 1991).

Humanistic perspectives in the science curriculum have a long history. Historical research in science education is summarized in Chapter 2, in which I pay particular attention to events occurring after World War II. The chapter provides a context for appreciating both the educationally and politically driven agendas that motivate the research found in the literature, and for understanding the literature's conceptualizations of a humanistic perspective in school science.

The book encompasses three forms of any curriculum: the *intended*, *taught*, and *learned* curriculum. An intended humanistic curriculum (Chapter 3) relates to curriculum policy that determines which humanistic perspectives are sanctioned and how those decisions are reached. The taught curriculum comprises the classroom materials that support humanistic science teaching (Chapter 4) and the teachers' orientations that determine what they will implement in their school science (Chapter 5). The learned curriculum, of course, is the content students actually learn, intended or not (Chapter 6).

Arising from discussions in Chapter 3 on the cultural relevance of school science, Chapter 7 addresses cross-cultural, multicultural, and high-poverty urban issues, including students' cultural self-identities and the integration of indigenous science (e.g., Native American or Aboriginal science) in the science curriculum, all in the context of humanistic school science.

In Chapters 3 to 7, pertinent research studies are synthesized giving emphasis to validated findings and research methods. In my last chapter, Chapter 8, I comment on fertile directions for further research into humanistic school science.

A humanistic perspective is not the only researched innovation to challenge the status quo of school science. Other innovations include constructivism (e.g., Millar, Leach, & Osborne, 2000), project-based learning (e.g., Moje, Callazo, Carrillo, & Marx, 2001), technology-design courses (e.g., Graube, Dyrenfurth, & Theuerkauf, 2003), environmental education (e.g., Kempton, Boster, & Hartley, 1995), gender-sensitive curriculum and instruction (e.g., Scantlebury, 1998), citizenship education in general (e.g.,

Davies, 2004), and science for political activism (e.g., Calabrese Barton, 2001a; Roth & Désautels, 2004; Roth & Lee, 2004). Researchers who champion any of these innovations can claim a connection to a humanistic perspective because their innovation includes one of the diverse meanings attributed to humanistic school science (Table 1.1). Their innovations, however, rest on more specific educational agendas than the general humanistic agenda that guides this book. Thus I do not treat these innovations per se as humanistic alternatives to a pipeline agenda, but their research projects do surface in this book to the extent that they illuminate features of a humanistic science program.

For example, a humanistic science curriculum is transformative in that it nurtures, at the very least, the development of a critical stance toward science and technology. Although transformative learning "involves experiencing a deep, structural shift in the basic premises of thought, feelings, and actions" (O'Sullivan, Morrell, & O'Connor, 2002, p. xvii), such learning can be mild or radical depending on the educational agenda of both teachers and students (O'Loughlin, 1992). This book reflects the research literature in science education, and thus I address a broad range of transformative agendas concerning student enlightenment and social change. In contrast, my colleagues working for political activism consistently advance a radical transformative agenda alone.

In-depth studies into students' abilities to deal with philosophical and social aspects of science suggest that overt humanistic content is more suitable for students aged 11 and older (Pedretti, 1999; Solomon, Duveen, Scot, & McCarthy, 1992), and aged 16 or older for controversial issues (Driver, Leach, Millar, & Scott, 1996). Accordingly, my synthesis of the literature restricts itself to school science that serves 11- to 18-year-olds, about grades 6 to 12 in North America.

My goal is to advance the ongoing debate over the purpose of school science and over the legitimacy of the traditional pipeline ideology represented in North America, for instance, by *Project 2061* (American Association for the Advancement of Science [AAAS] 1989) and *Standards* (National Research Council [NRC] 1996). I explore the evidence-based findings that shed light on school science practice and place those findings in a context of political reality. The book generally excludes nonresearch literature that simply advocates a position or offers a rationale for a humanistic perspective. In short, I eschew literature that articulates a perspective to be enacted by others, but I synthesize the research into that enacted perspective.

2

A Short History of Humanistic Perspectives in School Science

School subjects are grounded implicitly in the historical process and cultural conventions in which they arose (Sáez & Carretero, 2002). This chapter uncovers the historical process and cultural conventions that gave rise to the first school science curriculum in 19th-century England and the United States. At the same time, the chapter explores the reasons why the ideology underlying that first science curriculum overpowered an alternative ideology supportive of a humanistic science curriculum. To help clarify what a humanistic perspective means, I summarize historical research that illuminates both the history of the science curriculum and the most recent history of humanistic movements in science education.

The ideology of today's science curriculum is easily understood when placed in the historical context of its 19th-century origin, which occurred at a specific time within the evolution of science itself. Thus the historical processes and cultural conventions responsible for humanistic perspectives in school science are necessarily nested within the history of the science curriculum in general, which itself is nested within the history of science in the Western world. Accordingly, this chapter begins with a brief account of the development of the knowledge of nature held in esteem by Western cultures (i.e., Western science), but in a way that anticipates today's renaissance in the knowledge of nature held by indigenous peoples worldwide, a renaissance that is expanding our contemporary meaning of humanistic school science.

THE DEVELOPMENT OF THE SCIENCES

The accounts of the origin of what we today call science vary depending on one's cultural or academic perspective. Historians of ancient science, for instance, study the "science" (the systematic knowledge of nature) of

Babylonia, Egypt, India, China, Greece, and Islam (Dampier, 1971). Well before these ancient cultures flourished, however, other cultures existed in Paleolithic and Neolithic times. They possessed indigenous knowledge systems that described and explained phenomena of nature (i.e., "indigenous sciences"), systems entirely different from Babylonian science, for instance.

Within Neolithic cultures a dramatic innovation appeared in Asia Minor about 10,000 years ago: The agricultural revolution began. Some groups bought into it, some did not. As a result, from that time onward two fundamentally different types of cultures existed in the world. Each had its own culture-based stories that interrelated humans, the world, and the gods. These culture-based stories persist today in one form or another. Instead of labeling the two types of cultures with the conventional Eurocentric descriptors *civilized* and *primitive*, I borrow Quinn's (1992) more descriptive terms *Takers* and *Leavers*, respectively.

The Takers' cultural stories place humans above nature in hierarchical importance, and so when Takers live their lives (i.e., when they enact their cultural stories), it is common sense for them to take from nature; after all, the world is a human life-support system designed to sustain human life. The Leavers' cultural stories, on the other hand, tend to place humans equal to or below nature in hierarchical importance, and so when Leavers enact their cultural stories, it is common sense for them to leave nature or give back to nature in a way that disturbs nature as little as possible; after all, harmony with nature will sustain human life. Knowledge of nature developed by Takers will certainly differ in many ways compared with the knowledge of nature developed by Leavers. Babylonian, Egyptian, and Greek cultures honed the Takers' views of nature and communicated their own knowledge of nature in the written tradition. The Leavers, on the other hand, taught their youth in the oral tradition emphasizing the value of harmony with nature, not in a romantic sense attributed to their culture by some modern-day ecologists but, instead, in a survival sense (Knudtson & Suzuki, 1992).

The different knowledge systems of nature held by the Leavers and the Takers have evolved over the years into 21st-century indigenous sciences and into 21st-century Western science, respectively. These histories (described below) have influenced our conceptions of humanistic school science. For instance, today's cross-cultural science education (Chapter 7) encourages students to understand, but not necessarily believe (Cobern, 1996b), aspects of Western science with its powerfully predictive universal abstractions as well as to understand the local indigenous science that resonates with students' self-identities and has demonstrated either its powerful survival value over the ages (Battiste & Henderson, 2000; Cajete, 2000b; Kawagley, 1995; Semali & Kincheloe, 1999) or its current utilitarian value (Layton, 1986).

Indigenous Sciences

Indigenous cultures are found worldwide: for example, Native Americans; First Nations of Canada; Indian nations of South America; the Saami of Europe; indigenous peoples of Africa; the Aboriginal families of Australia; the Ainu of Japan; and Polynesian nations of the Pacific, including the Mäori of Aotearoa New Zealand and the first people of Hawai'i. These cultures possess systematic, empirical, rational knowledge systems of nature that serve their people in helpful ways. Their knowledge systems are defined as "indigenous" by Battiste and Henderson (2000) using international standards. Because an indigenous knowledge system is inseparable from the culture in which it evolved, some authors prefer the term *indigenous knowledge* to *indigenous science* (e.g., Semali & Kincheloe, 1999). The science education research literature, however, often identifies these indigenous knowledge systems by such phrases as Native science (Cajete, 2000b), Aboriginal science (Christie, 1991), Mäori science (McKinley, 1996), and Yupiaq science (Kawagley, 1995), to name just a few. Each represents one of many indigenous sciences. I use the term *indigenous sciences* because it seems more equivalent to the term *Western science* in the spirit of social equity.

In this book I follow Elder Beatrice Medicine's (2000) lead and refer to people belonging to any of the indigenous cultures worldwide as *Aboriginal*, with the acknowledgment that the term implies a great diversity of cultures rather than a homogeneous stereotype.

In the oral histories of many Aboriginal peoples, the original teachers of indigenous science were often godlike emissaries sent from the creator near the beginning of time to teach people how to survive in their land. In parts of Australia, for instance, Rainbow Serpent instructed the ancient people in central Australia at a special time called Tjukurpa (crudely translated, Dream Time), teaching them, for example, what to gather, when to gather it, and how to use it for survival. Similarly in North America (Turtle Island), the Pueblo people have Kokopelli and the Hopi nation has Sotuknang (Cajete, 2000b). Indigenous science content is usually held by certain elders, chosen when they were young to be trained over many years. This detailed knowledge can be communicated in a story or legend genre to others as circumstances dictate. Indigenous knowledge systems are constantly tested against the local environment. Consequently they are dynamic and tentative, changing to meet conditions that affect the community's survival (Cajete, 2000b; Riggs, 2005; Semali & Kincheloe, 1999; Snively & Corsiglia, 2001). More information is presented in Chapter 7.

Sometimes details in a recounted story have been modified through the influence of colonizing nations, including well-meaning, yet hegemonic, environmental enthusiasts. On Turtle Island (North America) this happened to a speech by Chief Seattle. It was rewritten entirely and made famous by a Texan environmentalist, and is now recognized as a hoax (Knudtson & Suzuki, 1992).

Today the world's indigenous sciences are often researched in such contexts as globalization, health, biopiracy, the destruction of biodiversity, agriculture, and healing from the effects of European and American colonization (Shiva, 2002). These contexts are described in a new journal, *Indilinga—African Journal of Indigenous Knowledge Systems*, published since 2002 in South Africa. The research into indigenous sciences is filled with tensions arising from a passion for self-governance, sovereignty, sustainable development, and cultural survival (Battiste & Henderson, 2000; McKinley, 2003; Menzies, Archibald, & Smith, 2004; Smith, 1999).

Western Science

I use the term *Western science* to draw our attention to the fact that the scientific enterprise associated with traditional school science evolved within Euro-American cultural settings (Pickering, 1992; Rashed, 1997), even though ideas were appropriated from other cultures, even though people from other cultures can participate in Western science, and even though not all Western people accept Western science as the most legitimate knowledge system. The evolution of Western science is succinctly summarized here to give context to the development of the first Anglophone school science curriculum.

Our Eurocentric line in the evolution of Western science often begins with the Greek origins of *philosophy*. Pure abstract ideas are associated with, for example, Thales, Pythagoras, Plato, Aristotle, and Ptolemy (Clagett, 1963; Dreyer, 1953). This Greek philosophy radically advanced in Western Europe during the 16th and 17th centuries (after the Renaissance period) with the establishment of *natural philosophy*, a new knowledge system based on the authority of empirical evidence and imbued with the value of gaining power and dominion over nature (Mendelsohn & Elkana, 1981). This historical advance is known today as the Scientific Revolution. Natural philosophy became a social institution when the Royal Society in England and l'Academie des Sciences in France were founded in the 1660s. Natural philosophers (e.g., Newton, Boyle, and Lavoisier) successfully inquired into the physical universe. Because Europeans were utilizing the printing press and colonizing the world at that time, natural philosophy was exported worldwide (along with Christianity), aiding the forces of colonization.

The success of natural philosophers at exercising power and dominion over nature attracted the attention of entrepreneurs who adapted the methods of natural philosophy to gain power and dominion over human productivity, in the context of various industries emerging across 18th-century Britain (Mendelsohn, 1976). This gave rise to the Industrial Revolution and provided a new social status for technologists. These industrialists spoke of natural philosophy as the handmaiden of technology. However, the independent-minded natural philosophers would have none of it. In the early 19th century, natural philosophers began to distance themselves from technologists, thereby precipitating the next radical transformation in the evolution toward modern Western science (Mendelsohn & Elkana, 1981).

Natural philosophers, led by Whewell (an Anglican priest and natural philosopher of mineralogy at Trinity College Cambridge), set about to revise the public image of natural philosophy by portraying technologists—for example, James Watt of steam engine fame—as people whose success depended upon *applying* the abstract knowledge of natural philosophy (Fuller, 1997; Layton, 1991). He and his colleagues succeeded in their revisionist project, and today there is a widespread belief in the erroneous notion that technology is solely applied science, thereby maintaining the ancient Greek ideology that holds "pure science" superior to practical technology (Collingridge, 1989).

Revising history was only one step in the 19th century's radical advance toward modern Western science. A new social institution was required and it needed to secure a social niche in 19th-century society. In short, natural philosophy needed to be *professionalized* (Layton, 1986; Mendelsohn, 1976; Orange, 1981). Very purposefully and politically, the name *science* was chosen to replace *natural philosophy* during the birth of a new organization in 1831, the British Association for the Advancement of Science (BAAS). "In seeking to achieve wider public support for science, the British Association wanted to present its members as a group of men united by a common dedication to the investigation of nature" (Yeo, 1981, p. 69). With the inauguration of the BAAS in 1831, a new meaning for *science* was added to the English lexicon, a meaning we primarily use today (Orange, 1981). In a speech to the 3rd annual meeting of the BAAS in 1834, Whewell coined the term *scientist* to refer to the cultivators of the new science—those who attended annual meetings of the BAAS (MacLeod, 1981). Interestingly, the term met with stiff resistance for half a century due to the popularity of the alternative "men of science."

To accommodate and organize participants at yearly BAAS meetings, themes for concurrent sessions were adopted, based on the administrative structure of the new University of Berlin, founded in 1810, which partitioned natural philosophy into the disciplines of physics, chemistry, geology, zoology, botany, and so on (Fuller, 1997). This classification scheme would eventually determine the structure of the science curriculum in the 1860s.

In addition to providing a professional identity for scientists, a professionalized science required the authority to decide who would be included and who would be excluded. This gatekeeping role was quickly taken up by universities where new disciplinary departments were established. By ensconcing Western science within the cloisters of university academia where it could control access to the various disciplines, and by defining what those disciplines would entail, the professionalization of natural philosophy into science was essentially complete in England by 1850.

The BAAS served as a model for the American Society of Geologists and Naturalists when, in 1848, the Society established the American Association for the Advancement of Science (AAAS, 2002). Similar to those of the BAAS, the prime functions of the AAAS were to promote the cultivation of science across the United States, to give systematic direction to scientific research, and to procure resources for its members.

Nineteenth-century science continued to evolve during the 20th century. World War II likely reshaped science more than any single historical event (Mendelsohn & Elkana, 1981). Abstract science was forced to cohabit with practical technology in order to defeat the Axis powers and preserve democracy. This unlikely marriage between science and technology produced a new social institution, research and development (R&D). Today only a small minority of academic scientists, less than 5%, undertakes purely curiosity-oriented research (Council of Science and Technology Institutes, 1993). The dominant patrons of R&D include business, industry, the military, government, and private foundations.

Events of the 20th century radically transformed 19th-century Western science into modern Western science, a process Ziman (1984) called the *collectivization* of science. Scientists and engineers still strive for power and dominion over nature, but mostly as R&D and in a new social context where technology, values, corporate profits, national security, and social accountability play an increasingly important role (Hurd, 1994; Solomon, 1994a, 1994b). Although 20th-century R&D essentially replaced 19th-century science, the phrase "R&D" has not replaced "science" in today's vernacular; and consequently in our schools we continue to have a science curriculum and not an R&D curriculum.

THE SCIENCE CURRICULUM

The history of a formal curriculum for school science in Anglophone countries dates back to the 19th century (DeBoer, 1991; Del Giorno, 1969; Hurd, 1991; Jenkins, 1985; Layton, 1973, 1981). In the 1850s, the British school curriculum was overcrowded with religious studies, the classics, grammar, languages, mathematics, and history. There was little room for new subjects such as the sciences. It would take the prestige and influence of the BAAS to change that.

The BAAS approved its "Scientific Education in Schools" report in 1867 (Layton, 1981). The BAAS promoted an ideology of "pure science," serving a self-interest in gaining members for the association and in obtaining research funds for those members. This also resonated well with the 19th-century progressive education movement's ideology that stressed mental training (Layton, 1981). "It seemed that chemistry and physics had been fashioned into effective instruments for both intellectual education *and* the production of embryonic scientists. A common thread had been devised to the twin ends of a liberal education and the advancement of science" (Layton, 1986, p. 115, emphasis in original). As a result, English education reformers in 1867 produced a science curriculum for secondary schools that marginalized practical utility and eschewed utilitarian issues and values related to everyday life. Instead this curriculum reflected the BAAS's newly achieved divide between science and technology and, at the same time, it reinforced social-class ideologies that favored the elite upper class (Seddon, 1991). Not all upper-class citizens embraced the ideology of scientific preprofessional training, however. Some associated it with manual labor (Waring, 1979). This backlash gave further impetus to emphasizing elitism in school science. Overall, the mental-training argument certainly helped squeeze the new science disciplines into an already crowded school curriculum.

The BAAS official position on education in 1867 distinguished between public understanding of science for the general education of a citizen and preprofessional training for future members of the BAAS (Layton, 1981). Preprofessional training served the scientific community's interests but also gave support to the progressive education movement by promising it the following outcome: "The scientific habit of mind [is] the principal benefit resulting from scientific training" (Layton, 1981, p. 194). These ideologies quickly became the status quo for the science curriculum and have not changed much today, in spite of the collectivization of science during the 20th century (Aikenhead, 1994c).

In the United States, organized curriculum development for high school science began in earnest during the 1890s in the context of a debate between advocates for citizen science (e.g., "Science of Society," in Spencer, 1859, p. 90) and those favoring an elite pre-professional training. The latter position was encouraged by events in England and by the appearance in the 1860s of German schools that specialized in teaching scientific disciplines (Jenkins, 1985). The AAAS was absent from this debate because of its pre-occupation with its own survival as an institution between 1861 and 1894.

Prior to the 1890s, the science curriculum in the United States consisted of assorted topics in astronomy, physiology, geology, natural philosophy, physics, chemistry, zoology, and botany (Del Giorno, 1969). In 1892, the National Education Association established the Committee of Ten, chaired by Charles Eliot, President of Harvard University (Hurd, 1991; Kliebard, 1979). Ideologically, Eliot championed mental training but opposed screening high school students for college admission as a central function of schools. The committee proposed four areas of high school study, a menu of programs much broader than the college admission requirements at the time. Eliot harbored an unbridled optimism about the intellectual capability of all students ("science for all" in today's vernacular). Accordingly, Eliot's Committee of Ten was unanimously against streaming (tracking) students by ability within their elective interests. These and other proposals by the committee drew strong criticism (Kliebard, 1979). As often happens in the heat of public debate, one's opponents make false accusations that sometimes stick like Velcro. Eliot's critics erroneously accused the committee of imposing college entrance expectations on the high school curriculum, a criticism that college science faculty then embraced as a committee recommendation (Hurd, 1991). Thus, in the aftermath of the debate over the report by the Committee of Ten, the U.S. science curriculum stressed both preprofessional and mental training for elite students. By 1910, America's status quo for school science mirrored England's.

By contemplating the historical origins of today's traditional science curriculum, we recognize that it is essentially a 19th-century curriculum in its educational intent and organization. In addition, we can now better appreciate the powerful ideologies that guide and sustain school science today, and therefore appreciate why students are socialized by their teachers into scientific disciplines. This standard socialization process causes most science teachers to teach in very similar ways toward very similar goals worldwide (Aikenhead, 1984; Cross, 1997a; Gallagher, 1998). These goals reflect the ideologies of preprofessional scientific training, of mental development through learning abstract concepts, of dismissing practical utilitarian outcomes, and of screening students for university entrance. These ideologies will challenge any attempt to reform school science into a subject that embraces a humanistic perspective (Fensham, 1992, 1998; Hughes, 2000).

During and ever since the science curriculum's formal inauguration in 1867 (England) and 1893 (United States), there have always been educators who promoted school science as a subject that connects with everyday society. Different eras have brought different social, economic, political, and educational forces to bear on reforming the science curriculum into a humanistic type of curriculum (Benjamin, 1939; DeBoer, 1991; Del Giorno, 1969; Dewey, 1916). For example, Donnelly (2002) reviewed British initiatives and quoted from a 1957 report by the Science Masters' Association (an elite group of science teachers at the time) that concluded, "If more emphasis is laid upon the cultural and humanistic sides of science education, more young people may favour science teaching as a career" (p. 553).

Hurd (1986, 1991) reviewed American attempts at this type of reform, mentioning the applied science and technology courses in the early 1900s, "viewed by scientists as an educational fad and were soon replaced by . . . simplified versions of . . . university science courses" (1991, p. 254). Hurd also described a 1920 U.S. Bureau of Education report, a 1928 AAAS committee report, a 1932 NSSE study, a 1945 Harvard report, and the 1983 *Nation at Risk* report. Hurd's historical research concluded that every committee and report criticized the science curriculum as being too narrow in vision, in subject matter, and in organization to relate science and technology to human, social, and economic affairs. "What the critics are seeking is a new and more viable contract between schooling and society, one in which science and technology are more closely tied to human affairs and social progress" (Hurd, 1989a, p. 2). Other historical studies by Layton (1973, 1991) and Solomon (1994a), as well as the case study research by Fensham (1998), have shown how innovative humanistic proposals threatened the social privilege and power that benefited an elite student enrolled in a traditional science curriculum.

Throughout the 20th century, attempts at reforming the traditional school curriculum into a humanistic one have largely been unsuccessful (Donnelly, 2002; Hurd, 1986, 1991; Klopfer, 1992; Layton, 1991; Walberg, 1991). This research finding directs our attention to the complex political power struggles that occur when educators challenge traditional school science at the policy level or at the classroom level (Chapters 3 and 5). The issue of political power is visited throughout this book.

RECENT HUMANISTIC SCIENCE CURRICULUM MOVEMENTS

Empirical research by science educators synthesized in this book is framed by several post–World War II humanistic conceptions of school science.

These conceptions arose from several currents of intellectual thought. As mentioned in Chapter 1, one major current of thought is associated with educationalists Dewey (1916) and Benjamin (1939), who inspired science educators to develop students' capacities to function as responsible, savvy participants in a world increasingly affected by science and technology. This current of thought originally led to teaching the history and philosophy of science in schools (Fensham, 1992; Matthews, 1994). But when science educators embraced social responsibility, engendered by environmentalism and by the emergent sociology of science, humanistic science teaching became identified with a movement called "science-technology-society," STS (Ziman, 1980). More recently, other currents of intellectual thought have been developed into research programs that enhance the meaning of humanistic perspectives in school science.

This chapter's mission is to clarify what a humanistic perspective means in school science. Here, I account for recent movements toward a humanistic science curriculum, giving greatest attention to the STS movement, followed by culture studies in science education, and then multicultural, urban, and poststructuralist science education.

The Science-Technology-Society Movement

Details of the history of STS are found elsewhere (Aikenhead, 2003; Bybee, 1993; Cheek, 1992; Cutcliffe, 1989; Solomon, 2003b; Solomon & Aikenhead, 1994; Yager, 1996) but can be summarized as follows.

Many proposals for a humanistic alternative to school science were inspired by university STS programs formally initiated in the late 1960s in the United States, United Kingdom, Australia, and The Netherlands. Some university programs taught history and philosophy of science that eschewed sociological perspectives on science. Others embraced sociology, economics, and politics, and gave themselves the label STS. These university STS programs responded to perceived crises in responsibility related to, for instance, nuclear arms, nuclear energy, many types of environmental degradation, population explosion, and emerging biotechnologies. Thus social responsibility for both scientist and citizen formed one of the major conceptions for a humanistic perspective in school science (Aikenhead, 1980; Bybee, 1993; Cross & Price, 1992, 2002; Kortland, 2001). At the University of Iowa, for instance, a societal issue-oriented science curriculum project evolved from the integration of social studies and science (Casteel & Yager, 1968) and later in Colorado (McConnell, 1982).

A second major conception for humanistic school science to emerge was the view of science often associated with Kuhn's (1962) *The Structure of Scientific Revolutions*. This analysis was viewed as a challenge to the positivism and realism inherent in traditional science courses (Abd-El-Khalick & Lederman, 2000; Kelly, Carlsen, & Cunningham, 1993).

Interest in humanistic content in the science curriculum enjoyed a renaissance at several universities after World War II. At Harvard, for instance, President Conant encouraged his faculty to give serious attention to the history, philosophy, and sociology of science, encouragement taken up at the time by young instructors such as Cohen, Kuhn, and Mendelsohn, respectively, and enhanced by physicist Holton. They influenced Ph.D. student Klopfer, who produced the *History of Science Cases* (Klopfer, 1969; Klopfer & Watson, 1957) and who thoroughly researched their impact in schools (Klopfer & Cooley, 1963). Similarly influenced was Gallagher, who in 1971 presciently articulated a blueprint for an STS science curriculum (echoed in Hurd's 1975 seminal publication) that rationalized teaching scientific concepts and processes embedded in the sociology/history/philosophy of science, relevant technology, and social issues (i.e., one teaches content, process, *and context*). Probably the most influential science education project to emerge from Harvard was the *Project Physics Course* (Holton, Rutherford, & Watson, 1970), a historical and philosophical perspective on physics aimed at increasing student enrollment in high school physics (Cheek, 2000; Holton, 2003; Walberg, 1991; Welch, 1973). It stimulated many other humanistic curriculum innovations worldwide (Aikenhead, 2003; Irwin, 2000).

A conceptual framework for STS was achieved through the integration of two broad academic fields: first, the interactions of science and scientists with social issues and institutions *external* to the scientific community, and second, the social interactions of scientists and their communal, epistemic, and ontological values *internal* to the scientific community (Aikenhead, 1994d; Ziman, 1984). A humanistic perspective incorporates both fields. In practice, however, some STS projects narrowly focused on just one of these domains.

Culture Studies in Science Education

Cultural anthropology (Eisenhart, 2001; Spindler, 1987) and postcolonial scholarship (Cajete, 1986; Kawagley, 1990) are two separate but interconnected currents of thought that have influenced the direction of science education research toward humanistic school science, mostly within the last decade. Details are found in Chapter 7, but a few key historical threads are summarized here.

Several science education research programs have been grounded in an anthropological definition of culture: "an ordered system of meaning and symbols, in terms of which social interaction takes place" (Geertz, 1973, quoted in Cobern, 1991, p. 31). Kearney's (1984) framework for world-view was brought into science education by Cobern (1991, 1993, 2000), who illuminated our understanding of students' difficulties and resistance to learning Western science. Research conducted by Phelan, Davidson, and Cao (1991) in high schools, by Lugones (1987) in her educational work world, and by Traweek (1992) with high-energy physicists in Japan was interpreted in terms of the *ease* with which people move from one culture or subculture to another, an idea introduced into science education by Costa (1995). Pickering (1992) articulated how Western science is a cultural entity in itself, a framework fruitful to Pomeroy's (1994) research that mapped out the early territory of culture studies in science education.

Several other key concepts and assumptions inform culture studies in science education. Students' meaningful understanding of the natural world is viewed as a cultural phenomenon (Spindler, 1987), and consequently the act of learning school science is perceived as *culture acquisition* (Wolcott, 1991). Here the word *culture* has the conventional meaning ascribed by Geertz in 1973 (quoted above from Cobern, 1991, p. 31). Other definitions exist in the research literature (e.g., Aikenhead, 1997; Atwater & Crockett, 2003; Eisenhart, 2001). If we view learning as an act of culture acquisition, we gain an intuitive, holistic, and rich appreciation of students' experiences in a science classroom (Costa, 1995). This cultural view is an extension of social constructivist theories of learning, such as the one held by Driver, Asoko, Leach, Mortimer, and Scott (1994): "Learning science in the classroom involves children entering a new community of discourse, a new culture" (p. 11). The cultural perspective embraced by culture studies in science education recognizes traditional school science as an attempt to transmit the culture of Western science to all students (Contreras & Lee, 1990; Hawkins & Pea, 1987).

But cultural transmission can be supportive or disruptive to students (Battiste, 2000; Cobern, 1993, 2000; Jegede, 1995; Lee, 1997). When the culture of science supports a student's worldview, the process is called *enculturation*. When the culture of science is generally at odds with a student's worldview, the resultant forced cultural transmission is called *assimilation*. However, if students decide that a scientific idea is useful to them for whatever reason, and if they incorporate that idea into their everyday thinking, then this act is called *acculturation*. These key terms are summarized here, drawing upon my earlier work (Aikenhead, 1996, 1997):

- *Enculturation*: Students learn the canonical content of school science, which harmonizes with their own worldviews, by incorporating that content into their personal way of viewing the world. Scientific thinking enhances their everyday thinking.

- *Assimilation*: Students learn the canonical content of school science, which clashes with their worldview in some way, by replacing or marginalizing their own ideas and values with scientific ones. Scientific thinking dominates their everyday thinking.
- *Acculturation*: Students borrow some canonical content from school science because of its current utilitarian value, and this content either replaces old ideas or is added to students' repertoires of ideas. Everyday thinking is a combination of commonsense thinking and some scientific thinking, often depending on the context.

Chapters 3, 6, and 7 elaborate on this description of student learning.

Research into cross-cultural science teaching in non-Western countries was reviewed in 1981 by Maddock and by others since then (e.g., Baker & Taylor, 1995; Contreras & Lee, 1990; George & Glasgow, 1988; Jegede, 1995; Krugly-Smolska, 1994; Ogunniyi, 1988; Okebukola & Jegede, 1990). These scholars concluded that students in non-Western countries generally face achievement and retention difficulties, and that schools lacked the flexibility to nurture students' personal growth in science education. Similar research conducted with Aboriginal students in Western countries came to identical conclusions (AAAS, 1977; Christie, 1991). My attempt to synthesize these emerging research programs with the STS movement culminated in an article "Science Education: Border Crossing into the Subculture of Science" (Aikenhead, 1996).

In the meantime, a postcolonial current of thought promoted self-determination of Aboriginal peoples (Battiste, 2000; Bishop & Glynn, 1999; Cajete, 1986; Kawagley, 1990; Smith, 1999) and was influenced in part by critical theory (e.g., Bourdieu & Passeron, 1977; Foucault, 1980) and by more advanced ideas in cultural anthropology applied to education (e.g., Battiste & Barman, 1995; Eisenhart, 2001). Attention to indigenous sciences and their advantageous role in school science for Aboriginal students were articulated (Aikenhead, 1997; Cajete, 1986; Cobern, 1996a; MacIvor, 1995; Nelson-Barber & Estrin, 1995; Snively, 1990).

Cross-cultural science education research (synthesized in the chapters that follow, particularly Chapter 7) has matured to the extent that Springer Press began publishing the *International Journal of Cultural Studies in Science Education* in 2006. This sketch of research developments in science education based on anthropology of education and postcolonial scholarship highlights a few of the historical movements that have influenced today's notion of humanistic school science.

Other Research Programs

Drawing upon critical theorists and guided by critical pedagogy (e.g., Giroux, 1992), the education goal of teaching for equity and social justice was highlighted in multicultural, urban, and other poststructuralist science education research programs (Calabrese Barton, 1998, 2002, 2004; Hines, 2003; Kincheloe, 1998; Roth & McGinn, 1998; Seiler, Tobin, & Sokolic, 2001; Tobin, 1997). Researchers seek to break down sociopolitical barriers of race, ethnicity, language, and class, for example, that separate home life from school life, particularly in high-poverty neighborhoods. These research programs enhance a humanistic perspective to the extent that they focus on the relevance of school science (Chapter 3) and the extent that they develop students' self-identities (Chapter 7), students' future contributions to society as citizens, and students' interest in making utilitarian meaning out of Western and/or indigenous sciences.

On the basis of social cognitive research (e.g., Wertsch, 1991), but clearly drawing upon critical pedagogy as well, science educators have developed a sociocognitive model of learning fruitful to science education research (O'Loughlin, 1992; Roth & Lee, 2004). The model addresses humanistic issues of social power and privilege in the classroom, and thus some of these research studies add to our understanding of humanistic school science. However, the model's narrow focus on social cognition (in spite of using the term *sociocultural*) misses the breadth of perspective inherent in culture studies in science education (Aikenhead, 2001a; Cobern & Aikenhead, 1998).

FEATURES OF THE HUMANISTIC SCIENCE MOVEMENT

Important features to a conceptual framework for humanistic school science have been articulated in the research literature:

- The degree to which a humanistic perspective supports or challenges a traditional positivist and realist view of Western science (Bingle & Gaskell, 1994).
- The degree to which a humanistic perspective advocates being aware of an issue, making a decision on an issue, changing (transforming) oneself as a result of learning about an issue, or taking personal or social action on an issue—a feature particularly salient to social responsibility (Bouillion & Gomez, 2001; Calabrese Barton, 1998; Cross & Price, 1992, 2002; Dahncke, 1996; Kozoll & Osborne, 2004; Pedretti & Hodson, 1995; Roth & Désautels, 2002, 2004; Rubba, 1987; Solomon, 1988b; Wiesenmayer & Rubba, 1999).

- The degree to which humanistic content is combined with traditional canonical science content (Aikenhead, 1994d; Bartholomew et al., 2004; Jeans, 1998; McClelland, 1998).
- The degree to which Western canonical science content is combined with (or replaced by) other sciences: citizen science, science-in-action (i.e., science content transformed by the context in which it is used), or indigenous sciences familiar to students in their everyday world (Aikenhead, 2001a; Cajete, 1999; Fourez, 1997; Jenkins, 1999; Koosimile, 2004; Kozoll & Osborne, 2004; Lawrence & Eisenhart, 2002; Malcolm, 2004; Roth & Lee, 2004; Ryder, 2001; Zacharia & Calabrese Barton, 2004). These other sciences can be found in a number of different contexts (Chapter 3), for example, personal health, community well-being, livelihood, or leisure.
- The degree to which the content, processes, and values of technology are integrated into a humanistic perspective (Cajas, 1999; Cheek, 1992, 2000; Fensham, 1988).
- The degree to which school science is itself *integrated*—the integration of scientific disciplines, and the integration of school science with other school subjects (Cajete, 2000b; Ratcliffe & Grace, 2003; Venville, Wallace, Rennie, & Malone, 2002).
- The degree to which instruction/assessment occurs in a community context away from the school (Aikenhead, 2002a; Calabrese Barton, 2001b; Kawagley, 1995; Pillay, 1996; Roth & Lee, 2004).
- The degree to which schooling is expected to reproduce the status quo or be an agent of equity and social justice in educational settings (Apple, 1996; Battiste, 2000; Calabrese Barton, 2002; Cross & Price, 1999; Hodson, 1994; Rodriguez, 2001).

Slogans for a humanistic perspective in the science curriculum, such as STS, can change from country to country and over time. In every era slogans have rallied support for fundamental changes to school science (Roberts, 1983). Today there are many slogans for humanistic school science worldwide, for instance, science-technology-society-environment (STSE), science for public understanding, science-technology-citizenship, citizen science, functional scientific literacy, public understanding of science, Bildung (in Germany), socio-scientific issues, and cross-cultural school science. These humanistic science perspectives are often seen as vehicles for achieving such goals as science for all and scientific literacy, and for improving marginalized students' participation and achievement in Western science and technology.

New science programs in schools are inevitably contentious. In the 1970s the history of science was rated "X" for science classrooms by stakeholders of the status quo, and then the sociology of science was rated "X" in the 1990s (Allchin, 2004). I anticipate that this book's humanistic school science will draw a similar rating.

Will history repeat itself in the 21st century by rejecting a humanistic perspective in the science curriculum, as predicted by Rutherford (1988) at the AAAS when he surmised of STS, "just one more passing fancy in education" (p. 126)? Or will the events of the past 50 years indelibly change school science? Only time will tell, of course. But Jenkins (2004) seems quite certain that events today favor curriculum change because academic stakeholders are losing political ground to the public at large, to scientists and engineers outside academia, and to innovative science teachers. "What is perhaps beyond doubt is that the school science curriculum has become contested territory to a degree that has not been evident since the mid-nineteenth century" (p. 244).

CONCLUSION

Four main conclusions seem warranted. First, humanistic perspectives found in science curricula and in research projects are deeply embedded in various cultures worldwide and have existed for at least 150 years. They are not new.

Second, just as science had to compete in the 1860s with the classics and religion to get a foothold in the English school curriculum, today a humanistic perspective must compete with the preprofessional training of elite students (moving through the pipeline) to earn a place in the school science curriculum. This reflects a competition between two ideologies: a humanistic perspective that promotes practical utility, human values, and a connectedness with societal events to achieve inclusiveness and a student orientation, versus a traditional perspective that promotes professional science associations, the rigors of mental training, and academic screening to achieve exclusiveness and a scientist orientation. Society in general did not reach a consensus on the latter position, but instead specific stakeholders politically achieved these goals within 19th-century British and North American society, establishing an ideology that became the status quo. Thus historical precedents and the ensuing politics of social-class privilege, not consensus, buttress the traditional science curriculum today (Carlone, 2003; Hodson, 1994; Seddon, 1991; Tobin, Seiler, & Walls, 1999). Therefore, curriculum change requires humanistic science educators to focus on *political* goals as well as educational goals.

Third, while 19th-century professionalized science has been dramatically transformed into 21st-century R&D through the collectivization of science (Ziman, 1984), school science has not similarly undergone any lasting dramatic reform. Fundamentally, school science is out of date—a conclusion based on educational soundness, not political reality. Politically, the traditional science curriculum is far from being out of date.

And fourth, humanistic perspectives fit a variety of overlapping and divergent curricular frameworks identified in the historical literature and found in the research studies synthesized throughout this book. Hence it does not make sense to delineate an iron-clad definition of humanistic school science based on this rich array of research programs. However, central to humanistic school science is its relevance to students, usually determined by students' cultural self-identities, students' future contributions to society as citizens, and students' interest in making personal utilitarian meaning out of various kinds of sciences—Western, citizen, or indigenous. Lemke's (2001) distillation of the fundamentals of a humanistic perspective is worth considering: "It means viewing science, science education, and research on science education as human social activities conducted within institutional and cultural frameworks" (p. 296).

To bring a modicum of structure to the wide diversity of humanistic perspectives and to ensure that all three forms of the science curriculum are addressed (the intended, the taught, and the learned curriculum), the chapters that follow are organized around the topics of: curriculum policy and relevance (Chapter 3), classroom materials (Chapter 4), teacher orientation and teacher education programs (Chapter 5), student learning (Chapter 6), and culture studies (Chapter 7).

3

Curriculum Policy

Policy research, in part, enhances our understanding of the values that underlie policy issues, including: Whose interests are being served? How did those practices emerge? and How are they sustained? Policy research also clarifies alternative policies, factors that influence a policy's effectiveness, and "the implementation and consequences of policy in concrete settings" (Collins, 2004, p. 2). Science curriculum policy entails the development of a principle of action to be adopted by a stakeholder group (usually a government or school district) and directed toward any aspect of teaching and learning in school science (Collins, 2004).

Chapter 2 identified key values that underlie either a traditional or humanistic school science policy, and it provided a historical context for understanding how one policy prevailed over the other, thereby establishing today's traditional science policy for schools. Humanistic school science was summarized by this policy statement: a student orientation that animates students' cultural self-identities, their future contributions to society as citizens, and their interest in making personal utilitarian meaning of scientific and technological knowledge. The motivation to promote a humanistic policy for school science has arisen, on the one hand, from persistent humanistic ideologies about the purpose of school science deeply embedded in our culture, and, on the other hand, from periodic and specific episodes of disappointment with the traditional science curriculum, usually expressed in terms of national crises (Klopfer & Champagne, 1990; van den Akker, 1998) or in terms of the underrepresentation of marginalized students in science and technology programs and related careers (AAAS, 1977; Calabrese Barton, 1998; McKinley, Stewart, & Richards, 2004).

This chapter investigates and synthesizes four areas of research concerning the educational soundness of a curriculum policy that supports humanistic perspectives in school science. These areas of research are major failures of the traditional science curriculum, people's success in learning and using science in nonschool contexts, the relevance of curriculum con-

24

tent, and the processes for formulating curriculum policy itself. Each area is discussed in turn. Collins's (2004) last point about policy research, "the implementation and consequences of a humanistic policy in concrete settings," is explored throughout Chapters 4 to 7.

MAJOR FAILURES OF THE TRADITIONAL
SCIENCE CURRICULUM

Deficiencies in the traditional science curriculum have been a cornerstone for arguments in support of a humanistic perspective. At least four major failures are documented in research studies.

Declining Student Enrollments

The first major failure concerns the chronic decline in student enrollment (Dekkers & Delaeter, 2001; Osborne & Collins, 2000; Welch & Walberg, 1967) due to students' disenchantment with school science (Fensham, 2004a; Hurd, 1989b; Lyons, 2003; Science Council of Canada [SCC], 1984; Ziman, 1980) or due to students' cultural self-identities conflicting with students' perceptions of science and technology (Brickhouse, Lowery, & Schultz, 2000; Brickhouse & Potter, 2001; Carlone, 2004; Cobern, 2000; McKinley, 2003; Medvitz, 1996). The Office of Educational Research and Improvement (1986) in the United States, for instance, reported that less than half of U.S. 7th-grade students and even fewer 11th-grade students thought science would be important, useful, or helpful in their adult lives. This failure of traditional school science threatens its primary goal: to produce knowledgeable people to go into careers in science, engineering, and related jobs.

It is instructive to examine "the pipeline" data from a 15-year longitudinal study, beginning in 1977, conducted by the U.S. Office of Technology Assessment (Frederick, 1991). Of the initial sample of 4 million grade 10 students, 18% expressed an interest in continuing toward university science and engineering courses. Of these interested students, 19% lost interest during high school (i.e., they moved out of the pipeline). Then, during university undergraduate programs, 39% of first-year science and engineering students lost interest—*twice* the proportion of high school students. These quantitative data support in-depth qualitative research which concluded that the problem of qualified students moving out of the pipeline resides much more with universities and colleges than with high schools (J. D. Lee, 2002; Tobias, 1990). Another substantial reduction in the pipeline population occurred between high school graduation and first-year university, a transition that showed a 42% loss in the number of students interested in pursuing science and engineering courses (Frederick, 1991; Sadler & Tai, 2001). These data are partly explained by an in-depth United Kingdom study which discovered that highly capable A-level science students, particularly young women and minority students, switched out of science as soon as they received their school science credentials, because the curriculum *discouraged* them from studying science further (Oxford University Department of Educational Studies, 1989). Similar results were obtained from international studies (Gardner, 1998).

Most research into students' views of the science curriculum concluded that it is socially sterile, impersonal, frustrating, intellectually boring, and/or dismissive of students' life-worlds and career goals (Cleaves, 2005; Hurd, 1989b; Lee & Roth, 2002; Lindahl, 2003; Osborne & Collins, 2001; Reiss, 2000, 2004; SCC, 1984). This perception prevails even for science-proficient students who enroll in senior science courses in high school (Lyons, 2003). The most cogent single force acting against enrollment was found to be the culture of school science itself.

One major reason for advocating humanistic content in school science has been to reverse this chronic loss of talented students (Eijkelhof & Lijnse, 1988; Holton, 2003; Ziman, 1980). Evidence suggests that humanistic perspectives in the science curriculum can improve the recruitment of students (Brush, 1979; Campbell et al., 1994; Osborne, Duschl, & Fairbrother, 2003; Solomon, 1994a; Welch & Rothman, 1968).

Discrimination and Cultural Alienation

A second major educational failure of traditional school science arises from various forms of discrimination. Students who belong to certain cultures, subcultures, or socioeconomic groups are significantly underrepresented in tertiary science and engineering programs and related careers (AAAS, 1977; Bianchini & Solomon, 2003; Costa, 1995; McKinley et al., 2004; Norman, Ault, Bentz, & Meskimen, 2001; Ogunniyi, 1996; Seymour, 1995). The issue is not simply student alienation but equity and social justice. Aboriginal students' meager participation in science and science-related occupations undermines their cultures' goals of sovereignty, survival, and sustainable resource stewardship (Chinn, 1999; MacIvor, 1995; McKinley, 2001, 2003; Snively & Corsiglia, 2001).

Parallel cases of equity and social justice are made for other visible minorities (e.g., Atwater & Crockett, 2003) and for economically depressed groups (e.g., Calabrese Barton, 1998, 2002; Gilbert & Yerrick, 2001). In their review of the research literature on underrepresented groups, Bianchini, Cavazos, and Helms (2000) concluded, "Women and ethnic minorities are far from having the same opportunities in science education as White men" (p. 516). Discriminatory goals of school science emerged unexpectedly from research that revealed systemic exclusion of adolescents outside the cultural power structures that sustain schooling and traditional school science (Tobin, Seiler, & Smith, 1999). Although humanistic school science holds promise for marginalized students, researchers have only recently begun to assess humanistic approaches to school science designed to ameliorate the underrepresentation of conventionally marginalized students (Hammond & Brandt, 2004), an issue addressed throughout this book, particularly in Chapter 7.

Dishonest and Mythical Images

A third major educational failure of the traditional science curriculum concerns the dishonest and mythical images about science and scientists conveyed by school science (Aikenhead, 1973; Allchin, 2004; Duschl, 1988; Gallagher, 1991; Gaskell, 1992; Kelly et al., 1993; Knain, 2001; Larochelle & Désautels, 1991; Lederman, 1992; McComas, 1998; Milne & Taylor, 1998). Three consequences result: Some science-proficient students (including those from underrepresented groups) lose interest in taking further science classes, some students become interested in science but for reasons based on dishonest and mythical images, and many students become citizens (some in key positions in government and industry) who make decisions predicated on myths about the nature and social aspects of the scientific enterprise. One reason for offering humanistic content in school science has been to correct these false ideas. Chapter 6 synthesizes research into this outcome.

Learning Difficulties

A fourth documented major failure dates back to the 1970s' research into student learning: *Most students tend not to learn science content meaningfully* (i.e., they do not integrate it into their everyday thinking) (Anderson & Helms, 2001; Cobern, Gibson, & Underwood, 1999; Driver, 1989; Gallagher, 1991; Hart, 2002; Lijnse, 1990; Louden & Wallace, 1994; Osborne, Duschl, et al., 2003; Shymansky et al., 1997). In other words, little enculturation, assimilation, or acculturation normally occurs (Chapter 2). This evidence-based conclusion is usually explained by the lack of relevance in school science (e.g., Fensham, 2000b) or by a lack of adequate pedagogy (e.g., Driver et al., 1994). Many research programs in science education have attempted in different ways to solve this lack of meaningful learning (Millar et al., 2000; Novak, 2002). However, even for students preparing for science-related careers (e.g., nursing), very few of them integrate science curriculum content into their own thinking when employed in science-rich workplaces, and this ability tends to be unrelated to their success at passing science courses (Aikenhead, 2005; Chin, Munby, Hutchinson, Taylor, & Clark, 2004; Cobern, 1993; Duggan & Gott, 2002; Lawrenz & Gray, 1995). A corpus of research suggests that learning canonical science content meaningfully is simply not achievable for the vast majority of students in the context of traditional school science (Aikenhead, 1996; Cobern & Aikenhead, 1998; Costa, 1995; Hennessy, 1993; Layton et al., 1993; Osborne, Duschl, et al., 2003; Shamos, 1995; Shapiro, 2004). As a result, alternative science curriculum policies with a humanistic perspective have been proposed to radically change the meaning of "science" in "school science," a controversial idea to be sure.

Playing Fatima's Rules

An important consequence to this fourth *educational* failure of the traditional science curriculum is the reaction of most students and many teachers to the *political* reality that science credentials must be obtained in high school or a student is screened out of postsecondary opportunities. Empirical evidence demonstrates how students and many teachers react to being placed in the political position of having to play school games to make it appear as if significant science learning has occurred even though it has not (Atwater, 1996; Bartholomew et al., 2004; Costa, 1997b; Griffard & Wandersee, 1999; Larson, 1995; Loughran & Derry, 1997; Meyer, 1998; Novak, 2004; Roth, Boutonné, McRobbie, & Lucas, 1999; Sutherland, 2005). The many rules to these school games are captured by the phrase "Fatima's rules," named after an articulate participant in Larson's (1995) study. Two simple rules are rote memorization and going through the motions of learning without being intellectually engaged. More sophisticated rules are found in the literature cited above. Playing Fatima's rules, rather than achieving meaningful learning, constitutes a significant *learned curriculum* for students and a ubiquitous *hidden curriculum* in school science (Aikenhead, 2000b). Playing Fatima's rules is a key concept that resurfaces throughout the book.

Playing Fatima's rules is an understandable coping strategy by students who feel their teacher is attempting to assimilate them into the culture of science and who want to resist such indoctrination. Any curriculum policy that inadvertently but predictably leads students and teachers to play Fatima's rules is a policy difficult to defend educationally, even though the policy flourishes for political reasons (Fensham, 1992). Dzama and Osborne (1999) consider this phenomenon a significant reason for the failure of school science to affect the industrial and economic outcomes expected of a school science program in developing nations.

LEARNING AND USING SCIENCE IN NONSCHOOL CONTEXTS

Although meaningful learning of canonical science is largely unattainable for many students in the context of the traditional science curriculum, it seems to be attained to some degree in nonschool contexts in which people are personally involved in a science-related everyday event or issue (Calabrese Barton & Yang, 2000; Dori & Tal, 2000; Goshorn, 1996; Michael, 1992; Rennie & Johnston, 2004; Roth & Désautels, 2004; Tytler, Duggan, & Gott, 2001b; Wynne, 1991). Thirty-one case studies of this type of research were reviewed by Ryder (2001), who firmly concluded that *when people need to communicate with experts and/or take action, they usually learn the science content required.* The academic language of science, for instance, is meaningfully acquired only when people have a personal reason for using it in an everyday context and are given sufficient time to master it.

Even though people seem to learn science content in their everyday world as required, this learning is not often the "pure science" (canonical content) transmitted by a traditional science curriculum. Research has produced one clear and consistent finding: *Most often canonical science content is not directly useable in science-related everyday situations,* for various reasons (Cajas, 1998; Chin et al., 2004; Furnham, 1992; Gott, Duggan, & Johnson, 1999; Hennessy, 1993; Jenkins, 1992; Layton, 1991; Layton et al., 1993; Ryder, 2001; Solomon, 1984; Wynne, 1991). In other words, the empirical evidence from this research contradicts scientists' and science teachers' hypothetical claims that science is directly applicable to a citizen's everyday life. What scientists and science teachers probably mean is that scientific concepts can be used to abstract meaning from an everyday event. The fact that this type of intellectual abstraction is relevant only to those who enjoy explaining everyday experiences this way (i.e., those who have a worldview that harmonizes with a worldview endemic to Western science; Aikenhead, 1996; Cobern, 2000; Cobern & Aikenhead, 1998; Lemmer, Lemmer, & Smit, 2003) attests to the reason most students perceive science as having no personal or social relevance. These students often see school science as a foreign culture (Calabrese Barton, 2002; Costa, 1995; Thijs & van den Berg, 1995).

How well do science teachers apply science content outside the classroom? Lawrenz and Gray (1995) found that preservice teachers with science degrees did not use science content to make meaning out of an everyday event for which canonical science content was directly relevant (i.e., providing proof to support their explanation of a morning haze) but instead used other content knowledge such as valuing authority.

This one research result, along with the 31 cases reviewed by Ryder (2001), can be explained by the discovery that canonical science content must be *transformed* (i.e., deconstructed and then reconstructed according to the idiosyncratic demands of the context) into knowledge very different in character from the canonical science in the science curriculum. This happens as one moves from canonical science content for explaining or describing to practical content for taking action—"science-in-action" or "citizen science" or "transformed science" (Fourez, 1997; Irwin, 1995; Jenkins, 1992, 1999; Layton, 1991; Mayoh & Knutton, 1997; Roth & Désautels, 2002; Roth & Lee, 2004; Tytler, Duggan & Gott, 2001a; Zacharia & Calabrese Barton, 2004). "This reworking of scientific knowledge is demanding, but necessary as socio-scientific issues are complex. It typically involves science from different sub-disciplines, knowledge from other social domains, and of course value judgements and social elements" (Kolstø, 2000, p. 659). If I replaced "socio-scientific issues" in this quotation with "medical events on a hospital surgical ward," I would paraphrase the evidence-based conclusion I established for nurses (Aikenhead, 2005).

When the science curriculum does not include the difficult process of transforming abstract canonical content into content for taking action, canonical science remains unusable outside of school for most students (Layton et al., 1993). And when students attempt to master unusable knowledge, most end up playing Fatima's rules.

This empirical evidence supports an educational policy that would alter the present meaning of school science. A humanistic science curriculum embraces a judicial balance between everyday action-oriented science content (e.g., citizen science and/or indigenous science; Cajete, 1999; Koosimile, 2004; Malcolm, 2004) *and* canonical science content. Researchers Lawrence and Eisenhart (2002) concluded, "Science educators and science education researchers are misguided not to be interested in the kinds of science that ordinary people use to make meaning and take action in their lives" (p. 187) (a type of indigenous science; Ogawa, 2004a).

Given these research studies that question the efficacy of teaching for meaningful learning in the context of the traditional science curriculum, there would seem to be little educational advantage for a teacher "to cover" the science curriculum. Instead, there seems to be greater advantage to teaching fewer canonical science concepts chosen for their relevance to a humanistic perspective (Eijkelhof, 1990; Häussler & Hoffmann, 2000; Kortland, 2001). The latter approach is supported by a plethora of research studies based on standardized achievement tests of canonical science. The studies showed no significant effect on students' scores when instruction time for the canonical content was reduced to make room for the history of science, the nature of science, or the social aspects of science; and, on occasion, students in a humanistic science course fared significantly better on achievement tests of canonical science than did their counterparts in a traditional science course (Chapter 6).

In summary, a recurring evidence-based criticism of traditional school science has been its lack of relevance for the everyday world (Gibbs & Fox, 1999; Millar & Osborne, 1998; Osborne & Collins, 2000; Reis & Galvão, 2004; Reiss, 2000), a problem dating back at least 150 years (Chapter 2). The issue of relevance is at the heart of humanistic science curricula.

RESEARCH ON RELEVANCE

Humanistic approaches to school science represent many different views on relevance (Bybee, 1993; Cheek, 1992; Fusco, 2001; Irwin, 1995; Kortland, 2001; Kumar & Chubin, 2000; Layton, 1986; MacIvor, 1995; Matthews, 1994; Millar, 2000; Solomon & Aikenhead, 1994; Yager, 1996). *Relevance* is certainly an ambiguous term. Mayoh and Knutton (1997) characterized it as having two dimensions: (1) *"Relevant to whom?* Pupils, parents, employers, politicians, teachers?" and (2) *"Relevant to what?* Everyday life, employment, further and higher education, being a citizen, leisure, students' existing ideas, being a 'scientist'?" (p. 849, emphasis in original). In the *educational* context of a humanistic science curriculum, the first question is invariably answered "relevant to pupils." (In a *political* context, however, the answer is much different—relevant to scientists.) The second question (Relevant to what?) leads to various meanings of relevance for curriculum policy (Keane & Malcolm, 2003). In this book, however, the multidimensional character of relevance is defined by a more political question (Fensham, 2000b; Häussler & Hoffmann, 2000; Roberts, 1988): *Who* decides what is relevant? Research into humanistic curriculum policies is reviewed here according to seven types of relevance, a scheme developed in part from Fensham's (2000b) views about *who* decides what is relevant. These seven heuristic categories are explored in turn (see Table 3.1 for a summary of the seven categories).

Wish-They-Knew Science

This type of relevance is typically embraced by academic scientists, education officials, and many science educators when asked, What knowledge is of most worth? (Driver et al., 1996; Fensham, 1992, 2000b; Shumba &

TABLE 3.1. Research on Relevance

Type of Relevance	Who Decides What Is Relevant?
Wish-They-Knew Science:	Academic scientists, education officials, many science teachers.
Need-to-Know Science:	The general public who have faced and resolved real-life problems/decisions related to science and technology.
Functional Science:	People in science-based occupations.
Enticed-to-Know Science:	The media and internet sites.
Have-Cause-to-Know Science:	Experts who have interacted with the general public on real-life issues.
Personal-Curiosity Science:	Students themselves.
Science-as-Culture:	Interpreters of culture who can determine what aspects of science comprise features of a local, national, and global culture.

Glass, 1994; Walberg, 1991). The usual answer, canonical science content, moves students through the pipeline for success in university programs. Even when teachers choose everyday contexts to teach wish-they-knew science, their ideas of a relevant everyday context are often at odds with most students' views (Campbell et al., 1994; Gagné, 2003; Lubben & Campbell, 1996).

But exactly how relevant is this wish-they-knew content for science-oriented students in first-year university courses? Research suggests it is not as relevant as one might assume and, on occasion, not relevant at all (Aikenhead, 1994b; Champagne & Klopfer, 1982; Mason, 1996; Yager & Krajcik, 1989; Yager, Snider, & Krajcik, 1988). First-year university students who had not studied the prerequisite physical science course in high school achieved as well as their counterparts who had enrolled in the prerequisite. Graduates of a high school humanistic chemistry course, *ChemCom*, who enrolled in a first-year university chemistry course for nonmajors fared as well as their counterparts who had studied traditional, honors, or advanced placement high school chemistry. Sadler and Tai's (2001) survey research claims, "Taking a high school physics course has a modestly positive relationship with the grade earned in introductory college physics" (p. 111). An endorsement of "modestly positive" would seem to be faint praise indeed. These research studies should rationally assuage science teachers' fear that time spent on humanistic content will diminish students' chances of success at university.

Although the *educational arguments* favoring wish-they-knew science are particularly weak, *political realities* favoring it are overwhelming (Fensham, 1993, 1998; Gaskell, 2003; Hart, 2004).

An alternative approach to establishing wish-they-knew science comes from continental Europe where the tradition called "didaktik" guides teachers in their analysis of potential canonical science content according to diverse criteria of relevance related to the discipline taught and to the purposes of their school (Fensham, 2004b). Didaktik analysis represents a different cultural tradition to curriculum development. Traditions vary worldwide.

Need-to-Know Science

This type of relevance is defined by the lay public who have faced a real-life decision related to science, exemplified by the Science for Specific Social Purposes project (Layton, 1991; Layton et al., 1993), a study of parents dealing with the birth of a Down's syndrome child, old people dealing with energy use, workers at a nuclear power plant dealing with scientific information on radiation effects, and town councilors dealing with the problem of methane generation at a landfill site. Curriculum policy researchers asked, What science content was helpful to them in making their decisions? Ryder's (2001) answer, based on his analysis of 31 case studies of need-to-know science, was the same as the conclusion I reached 16 years earlier, based on a more modest analysis (Aikenhead, 1985). Ryder wrote, "Much of the science knowledge relevant to individuals in the case studies was *knowledge about science*, i.e. knowledge about the development and use of scientific knowledge rather than scientific knowledge itself" (p. 35, emphasis in original).

In other words, for "science" in "school science" to be relevant, the curriculum's definition of science content must expand to include knowledge *about* science and scientists, defined in Chapter 1 as one way to characterize humanistic content.

One reason that people tend not to use canonical science content in their everyday world (in addition to its not being directly useable, as described above) is quite simple: Canonical science content is the wrong type of content to use in most socioscientific settings. Need-to-know science (e.g., citizen science and knowledge *about* science and scientists) turns out to have greater practical value than canonical science—an issue explored in greater detail in Chapter 6.

Even though need-to-know science has potential for rationalizing the selection of specific humanistic content for the science curriculum, its potential may be limited. Fensham (2000a) suggested the following reason: "Its retrospective unpredictability, its variation of experience among citizens, and the time gap between school and the 'need,' make it unattractive to curriculum designers of school science" (p. 74).

Functional Science

Functional science is deemed relevant primarily by people with occupations or careers in science-based industries and professions. Coles (1998) surveyed higher education specialists in science and United Kingdom employers to identify scientific content thought to be essential to school science. Unexpectedly the content found in the traditional science curriculum (wish-they-knew science) received very limited consensus across all respondents. The most valued qualifications were generic thinking skills and mathematical capabilities. Large United Kingdom organizations, such as the Chemical Industries Association and the Association for the British Pharmaceutical Industry, preferred their recruits to possess general capabilities rather than specific canonical science content (Coles, 1996). Desired capabilities included commitment and interest; skills in communication, numeracy, and information technology; personal effectiveness, relationships, and teamwork; and initiative and creativity. Of lesser importance was the list of scientific capabilities sought by these employers (from highest to lowest priority): decision making by weighing evidence; scientific skepticism and logical thinking; and, finally, understanding science ideas.

Coles's (1998) results with higher education specialists were replicated in Beijing, China (Law, 2002), where leading scientists were interviewed to find out their preferred qualities of a recruit for their area of expertise. All scientists mentioned creativity, and several talked about a person's personal interest, perseverance, willingness to inquire, and ability to communicate. The results of this in-depth study surprised the researchers:

> Contrary to much conventional wisdom and the practice of specifying lists of content as the key elements of a science curriculum, the study found that habits of mind, attitudes, and values figured more prominently than scientific content knowledge in the attributes identified by "societal experts" as the most distinguishing qualities for success in their particular areas of expertise. (p. 151)

The experts relegated "having a broad or wide knowledge base" (p. 158) to a relatively low priority.

In short, key characteristics of functional science include mathematical capabilities, thinking skills, personal attributes related to acquiring science content, and personal skills required to communicate science content. Knowing specific canonical science is a low priority.

But we often hear that a country's economic development depends on a workforce strong in its scientific conceptual knowledge (Walberg, 1991). However, economic studies have consistently shown that economic development depends on factors beyond the influence of school science, for example, emerging technologies, industrial restructuring, poor management decisions, and government policies that affect military development, monetary exchange rates, wages, and licensing agreements (Bishop, 1995; Cuban, 1994; David, 1995; Drori, 1998; Halsey, Lauder, Brown, & Wells, 1997). Walberg (1991) got it wrong (Solomon, 1997).

Why did industry personnel place "understanding science ideas" at the lowest priority for judging a recruit? The answer comes from the ethnographic research by Duggan and Gott (2002) in the United Kingdom, Law (2002) in China, and Lottero-Perdue and Brickhouse (2002) in the United States. Their *on the job* interviews indicated that the science content used by science graduates at work was so context specific it had to be learned on the job. High school and university science content was rarely drawn upon.

On the other hand, Duggan and Gott's (2002) data suggested that procedural understanding (i.e., the thinking directly related to doing science-like tasks) was essential across most science-related occupations. Specifically Duggan and Gott discovered one conceptual domain that was applied by all workers: "concepts of evidence." Concepts of evidence were used to critically evaluate scientific data, for instance, concepts related to the validity and reliability of data, and concepts of causation and correlation. Similar findings arose in their research with an attentive public involved in a science-related societal issue. Duggan and Gott spoke for many researchers (e.g., Fensham, 2000a; Ryder, 2001) when they concluded, "Science curricula cannot expect to keep up to date with all aspects of science but can only aspire to teach students how to access and critically evaluate such knowledge" (p. 675).

The humanistic perspective germane here concerns a correct understanding of concepts of evidence when dealing with social implications; for instance, Is the scientific evidence good enough to warrant the social action proposed? In this context it is useful to understand the ways in which scientific evidence is technically and socially constructed (Allchin, 2004; Bingle & Gaskell, 1994; Cunningham, 1998; Kelly et al., 1993; McGinn &

Roth, 1999), and thus it would be useful to include this humanistic content in the science curriculum because it speaks directly to critically evaluating scientific evidence.

An important capability valued by employers and employees in science-based occupations was the capability *to learn* science. The implication for curriculum policy is to include canonical science content in school science but to recognize it as being secondary in importance compared to objectives more directly related to a humanistic perspective in the science curriculum. For example, learning to critically analyze scientific evidence requires scientific concepts to be sure, but it matters little which canonical concepts appear in a curriculum as long as they are pertinent to the evidence at hand (Aikenhead, 2005; Kolstø, 2001b; Law, 2002; Ratcliffe, 1999; Symington & Tytler, 2004).

In a project that placed high school students into science-rich workplaces (e.g., veterinary clinics and dental offices), Chin et al. (2004) investigated ethnographically (1) the relationship between school science and workplace science (a type of functional science) and (2) the participants' perceptions of that relationship. The fact that students saw little or no connection between school science and workplace science was explained by the researchers this way: Knowing canonical science content was not relevant to one's accountability in a science-rich workplace (a conclusion very similar to the results of research into need-to-know science, discussed above). Chin et al. concluded that workplace science (functional science) met the purpose and accountability of the workplace, causing workplace science to differ qualitatively from school science. At the same time, the students in the study were accountable to their school science program, but they fulfilled this responsibility by playing Fatima's rules.

A number of ethnographic research studies in urban science education demonstrated the success of connecting students to the work world by involving members of the community in teaching both functional science and wish-they-knew science (Calabrese Barton, 2001b, 2002, 2004; Calabrese Barton & Yang, 2000; Hammond & Brandt, 2004). For instance, one gardening project afforded authentic activities for inner-city youth ($N = 23$) guided by four adult leaders and two master gardeners (Rahm, 2002). By learning to grow marketable crops, students accrued profits which they invested by physically improving their neighborhood. The science that students learned emerged from collaborative training for specific jobs. Such projects can also include indigenous science (Hammond, 2001) and computer technology skills found in the United States–Africa project Science, Technology, and Culture (Kyle, 2004).

A very different avenue of research was pursued by Häussler and Hoffmann (2000) in Germany. They showed how the Delphi research technique could establish an educationally sound curriculum policy for high school physics by getting diverse stakeholders to reach a consensus over the following question: "What should physics education look like so it is suitable for someone living in our society as it is today and as it will be tomorrow?" (p. 691). Their 73 participants represented stakeholders associated with wish-they-knew science (e.g., physicists and physics teachers) and with functional science (e.g., personnel officers in physics-related industries and general educationalists). Häussler and Hoffmann did not initially group their stakeholders into these two categories but instead used a hierarchical cluster analysis statistic to tease out like-minded stakeholders. This analysis produced two coherent groups: Group 1 favored "scientific knowledge and methods as mental tools" and "passing on scientific knowledge to the next generation" significantly more than did Group 2, who favored "physics as a vehicle to promote practical competence" (p. 693). These statistical results lend credence to the two categories of relevance that distinguish between wish-they-knew science and functional science. Interestingly, however, Häussler and Hoffmann found that both groups gave *highest* priority to topics related to "physics as a socio-economic enterprise" that show "physics more as a human enterprise and less as a body of knowledge and procedures" (p. 704).

Functional science usually lies outside the sphere of canonical science content normally transmitted in traditional school science. Functional science, however, can be found in progressive vocational science programs, although when vocational school teachers are concerned only with the narrow objective of vocational technology, vocational science loses its humanistic perspective.

Enticed-to-Know Science

By its very nature, enticed-to-know science excels at its motivational value. This is science content encountered in the mass media and the Internet— positive or negative in its images of science, and sensational and sometimes dishonest owing to the media's quest to entice a reader or viewer to pay closer attention.

Fensham (2000a) reports that the OECD's Performance Indicators of Student Achievement project is using enticed-to-know science "to see how well their science curricula are equipping [15-year-old] students to discern, understand, and critique the reporting of science in newspapers and the Internet" (p. 75).

Millar (2000) in the United Kingdom and Dimopoulos and Koulaidis (2003) in Greece described how their yearlong analyses of the content of science-related articles in their respective national newspapers identified the science and technology knowledge that would be most useful in making sense of the articles. Millar's analysis stimulated a revision of the AS-level syllabus in the United Kingdom and eventually culminated in Hunt and Millar's (2000) high school textbook *AS Science for Public Understanding*, which provides a humanistic perspective for students. For highly controversial issues, however, Thomas (2000) cautions policy makers over the extent to which "sound science" can be taught strictly from newspaper articles.

The "event-centered learning" approach to humanistic school science researched in Brazil and the United Kingdom (Watts, Alsop, Zylbersztajn, & de Silva, 1997) is another type of enticed-to-know science. Some classroom tasks require students to generate their own media and Internet material (e.g., preparing a television script).

Moral issues and public risk are often associated with enticed-to-know science because the media normally attend to those aspects of events (Cross & Price, 1992, 2002; Eijkelhof, 1990; Nelkin, 1995; Osborne, Duschl, et al., 2003; Sadler, 2004; Stocklmayer, Gore, & Bryant, 2001). The more important everyday events in which citizens encounter science and technology involve risk and environmental threats (Irwin, 1995).

Have-Cause-to-Know Science

This is science content suggested by experts who interact with the general public on real-life matters pertaining to science and technology, and who know the problems the public encounters when dealing with these experts (Law, Fensham, Li, & Wei, 2000). The process assumes that societal experts are better situated than academic scientists to decide what knowledge is worth knowing in today's changing scientific and technological world.

This empirical approach to developing curriculum policy was tested in China to determine the have-cause-to-know science for two different curricula, one aimed at everyday coping and the other aimed at socio-scientific decision making (Law, 2002; Law et al., 2000). To generate curriculum policy for everyday coping by the lay public (the first curriculum), four societal experts were drawn from the following four domains: home and workplace safety; medical, health, and hygiene problems; nutrition and dietary habits; and "consumer wiseness." These experts agreed that the public had cause to know basic scientific knowledge related to an event with which people were trying to cope and to know specific applications of that knowledge, but particularly to be able to critically evaluate cultural practices, personal habits, media information, and multiple sources of (conflicting) information (Law, 2002). During their interviews, the experts noted public misconceptions, superstitions, and cultural habits detrimental to everyday coping.

To generate curriculum policy for the second curriculum, the lay public's participation in socioscientific decision making, four experts were selected from Hong Kong's democratic institutions (the legislature, a government planning department, and a civilian environmental advocacy group) and were interviewed. The research showed that the public's have-cause-to-know science was very similar to that required of everyday coping, except socioscientific decision making required a broader knowledge base and more complex skills to critically evaluate information and potential solutions (Law, 2002). The researchers acknowledged the fact that decisions often rely more on applying values than on applying science content (a topic discussed in detail in Chapter 6 in the context of research on students' learning to make decisions). Overall, the Chinese experts placed emphasis on a citizen's capability to undertake self-directed learning (lifelong learning) but placed low value on a citizen's knowing particular content from the traditional science curriculum, a result reminiscent of research related to functional science (reviewed earlier). Although Law conducted her research on a small scale using a limited number of participant experts, her research design could be scaled up in other jurisdictions to create a substantial curriculum policy study for school science.

Fensham (2002) envisioned a have-cause-to-know science curriculum policy unfolding in three phases: (1) Selected societal experts systematically determine features of society endemic to an informed citizenry; (2) academic scientists specify science content associated with the features of society identified in phase 1; and (3) on the basis of the first two phases, science educators develop a school science curriculum.

Have-cause-to-know science is a feature of the Science Education for Public Understanding Project, SEPUP, in the United States (Thier & Nagle, 1994, 1996). Societal experts in industry, the sciences, and education provided the curriculum developers with elements of a relevant issues-based curriculum that led to STS chemistry modules and three STS textbooks (SEPUP, 2003).

In The Netherlands, Eijkelhof (1990, 1994) used the Delphi research technique to gain a consensus among *societal experts* to establish the humanistic and canonical science content for an STS physics module, "Ionizing Radiation." The 35 Delphi participants in Eijkelhof's study were carefully selected to represent a variety of fields and opinions on the risks of ionizing radiation. After the normal three rounds in the Delphi procedure, Eijkelhof's radiation experts pointed to suitable societal contexts of application and concomitant scientific content that the public had cause to know (Eijkelhof, Klaassen, Lijnse, & Scholte, 1990). This method of policy development bypassed the academic scientist, phase 2 in Fensham's scheme (just above). Although this may make sense educationally, it could be a poor tactic politically in some jurisdictions. Eijkelhof (1990) noted that policy research by itself should not *prescribe* the final curriculum. A curriculum development team must consider educational issues, for example, learning difficulties of students, available instruction time, and pedagogical factors (Fensham's phase 3). A curriculum team could attend to those issues by drawing upon science education research reviewed and synthesized in this book.

In contrast to the successful Dutch experience, an Australian chemistry curriculum committee could not reach a consensus on a balance between societal contexts of application and scientific content. As a result, the committee's writers tended to promote the status quo (wish-they-knew science) rather than the intended have-cause-to-know science (Fensham & Corrigan, 1994).

The National Curriculum in the United Kingdom calls for humanistic content to be taught but does not specify the content in any detail. Osborne, Collins, Ratcliffe, Millar, and Duschl (2003) employed the Delphi technique to establish a consensus in the United Kingdom on what "ideas about science" (humanistic content) should be taught in school science. During their three rounds of the Delphi procedure, 23 experts (professional and academic people notable for their contributions to clarifying science for the public) produced nine ideas that showed sufficient stability and support ("scientific methods and critical testing," "creativity," and "historical development of scientific knowledge," to name the top three). These research results informed the development of teaching materials that explicitly taught the nine ideas about science (Bartholomew et al., 2004). The have-cause-to-know ideas about science addressed humanistic content only; the canonical science content had been established by the National Curriculum's wish-they-knew science.

A disadvantage of the Delphi procedure is evident in ambiguous and "motherhood" statements that sometimes emerge. This is likely the result of participants' not meeting face to face to clarify the meaning of each statement.

Curriculum policy research has also included surveys of experts to determine which social issues (and, therefore, which have-cause-to-know science) they valued most in a humanistic science curriculum. Experts have included scientists and engineers, citizens, science teachers, and science educators in the United States and internationally (Bybee, 1993). The relevant contexts for have-cause-to-know science were identified, but their actual influence on curriculum policy has not been noticeable (Cheek, 2000). This survey research was perhaps more politically successful at raising awareness of STS than at developing specific curriculum policies.

Personal-Curiosity Science

When students themselves decide on the topics for school science, relevance takes on a personal, though perhaps idiosyncratic, meaning, as students' hearts and minds are captured (Gardner, 1998; Hodson, 1994; Osborne & Collins, 2000; Reiss, 2000). On the basis of a humanistic curriculum policy principle that one builds on the interests and experiences of students, Sjøberg (2000) surveyed over 9,000 13-year-old students in 21 countries to learn their past experiences related to science, their curiosity toward certain science topics, their attitude to science, their perception of scientists at work, and their self-identity as a future scientist. Based on the same curriculum policy principle, Häussler and Hoffmann (2000) surveyed over 6,000 German students, aged 11 to 16 years, to determine, among other things, their interest in various physics topics (i.e., the everyday context for the topic and its relevant content), their interest in physics as currently taught in their school, and their personal background factors. Data from Sjøberg (2000) and from Häussler and Hoffmann (2000) offer insights into students' differential interests; for instance, "music" was much more interesting than "acoustics and sounds," and "the rainbow and sunsets" much more so than "light and optics." In short, concrete themes embedded in student experiences were much more relevant than science discipline topics, a finding supported by three decades of research by the Dutch PLON project (Kortland, 2001).

In Sjøberg's (2000) study, students in non-Western countries had a significantly more positive (humanistic) image of scientists (i.e., heroic figures helping the poor and underprivileged) than did their counterparts in Western countries, a finding that points to the importance of culture in a student's everyday world (an issue addressed in the next subsection).

In the Häussler and Hoffmann (2000) Delphi study, students' views were congruent with those of stakeholders who advocated a humanistic perspective in the physics curriculum but discordant with the traditional physics courses offered at their school. Häussler and Hoffmann pointed out that a curriculum policy founded on the Delphi results would look very similar to a curriculum policy founded on student interests alone (i.e., personal-curiosity science). (As described in Chapter 6, this humanistic physics course did result in significantly greater achievement on canonical science content than did a traditional physics course.)

Sjøberg (2003) initiated an extensive international survey study of personal-curiosity science, the Relevance of Science Education (ROSE) project. The results and findings from many countries will contribute greatly to our understanding of personal-curiosity science and its cultural contexts (e.g., Ogawa, 2004b).

Surveys of student interest have typically accompanied the evaluation of humanistic science pilot materials. This research in North America produced fairly consistent results: The personal-curiosity science in which students expressed most interest was related to sex and drugs (Aikenhead, 1992; Stoker & Thompson, 1969). These preferences, however, shifted to problems of population and pollution when a different question was posed. For example, Stoker and Thompson also asked, What topics are of most value to you now? or What topics will be most valuable to you in the long run? In the United Kingdom, Ratcliffe and Grace (2003) asked, "What three things generally worry you?" and "What three things would worry you if they were changed or stopped?" (p. 152). The researchers reported data that showed how students' sex and age influenced their responses. One research concern underpinning these studies is quite simple: The context in which the survey is administered and the wording of its questions influence students' responses.

An alternative to survey research was illustrated by Malcolm (2004) when he and a colleague spent time in a rural Zulu community in South Africa to talk with students about what natural phenomena interested them. This research led to a curriculum policy to treat rain as a major science topic and to include Zulu science (an indigenous science) in the science curriculum. The development of appropriate teaching materials became a follow-up project, described in Chapter 4.

Science-as-Culture

A more holistic yet abstract concept of relevance for school science was advanced by Weinstein's (1998) research concerning the enculturation of students *into everyday society*, an approach to science education that stands in stark contrast to the status quo of traditional school science: the enculturation of students *into scientific disciplines*. (The distinction is discussed later in the section "Cultural Relevance.") Culture decides, de facto, what is relevant for science-as-culture (e.g., the role of "guinea pigs" in research on human beings). In the culture of school, "Students constantly are being measured, sorted, and turned into objects of scrutiny. They learn science up close and personal but not as scientists; rather, they learn it as objects of science" (p. 493).

Weinstein (1998) identified a network of communities (webs of scientific practice) in students' everyday lives: health systems, political systems, the media, environmental groups, and industry, to name a few. Each community interacts with communities of science professionals, resulting in a *cultural commonsense notion of Western science* described by Weinstein as follows:

> The meaning making that we call science happens in a way that is distributed over the society spatially and temporally. It happens through science fiction, it happens through laboratory work, ... it happens in hospitals, it happens in advertising, and it happens in schools. To emphasize this, I explicitly refer to *science-as-culture* rather than to just *science*. I do this as a reminder to the reader that I am concerned with science in all parts of the network and not just the laboratory, field station, and research institute. (p. 492, emphasis in original)

Science-as-culture is much more than just pop culture (Solomon, 1998).

As a category of relevance, science-as-culture serves in part as a superordinate category to need-to-know, functional, enticed-to-know, have-cause-to-know, and personal-curiosity science. Its relevance resides in a student's community culture (a commonsense notion of Western science) and in a student's home and peer cultures (Costa, 1995; Kyle, 1995; McKinley, 2005; Solomon, 1994c, 1999a, 2003a). Science's role in society is also embedded in science-as-culture, evidenced by such roles as setting standards, regulating commerce, providing legal evidence, announcing medical breakthroughs, creating novel ethical dilemmas, and requiring financial support for research and development (Dhingra, 2003; Jenkins, 2000; Pedretti, 2004; Stocklmayer et al., 2001).

Research into science-as-culture reveals useful ideas for a humanistic science policy, particularly for the enculturation of students into their local, national, and global communities. Prelle and Solomon (1996), for instance, provide a rich account of the differences between students' orientation to an environmental issue and their scientific knowledge on the subject. The researchers explored students' science-as-culture by investigating those differences in three settings: the science classroom, students' homes, and on holidays.

Science museums and science centers have presented the public with issues-based exhibitions that personalize humanistic science content, evoke emotions, stimulate discussions of that humanistic content, and promote self-reflection on the issue presented. In her review of 10 years of research into those types of exhibits, Pedretti (2004) concluded, "Science then is framed by culture—its context has social, economic, political, and historical dimensions" (p. S35). These dimensions are webs of scientific practice. Thus the content is science-as-culture.

In their research into scientific content in television commercials, McSharry and Jones (2002) posed a new curriculum policy research question: What scientific understanding is of critical value to consumers of television commercials? Nelkin's (1995) seminal research into science and the media raises an even broader policy question: What understandings of Western science and journalism are of critical value to consumers of the mass media?

Curriculum policy was investigated by ethnographic researchers Symington and Tytler (2004), who interviewed a stratified random sample of citizens listed in *Who's Who in Australia*, a diverse group of successful people who credibly represented a cultural commonsense notion of Western science (science-as-culture). The researchers did not complete the 20 interviews intended but stopped after 15 interviews because "theoretical saturation" had been achieved. These prominent citizens opined a purpose for school science that was coherently *humanistic*, giving emphasis to lifelong learning and rejecting knowledge accumulation they associated with their own school science experiences.

Science-as-culture can also be captured by project-based learning in which science-related real-life problems are addressed by students in an interdisciplinary way (Bouillion & Gomez, 2001; Calabrese Barton & Yang, 2000; Dori & Tal, 2000; Lee & Roth, 2002; Pillay, 1996; Roth & Désautels, 2002). This approach draws upon community resources and local culture to stimulate a combination of need-to-know, functional, and have-cause-to-know science (May, 1999)—in other words, science-as-culture. The presence of a humanistic perspective in a project-based curriculum depends, however, on the degree to which its humanistic content is made explicit in the instruction and assessment of students (Aikenhead, 1973; Kortland, 2001; Ratcliffe, 1997b).

Conclusion

These seven heuristic categories of relevance, based on who decides what is relevant, can help describe the content and contexts for a humanistic perspective of a particular science curriculum. More often than not, a curriculum will embrace several categories simultaneously by combining some wish-they-knew science found in a government curriculum document with, for example, functional science, enticed-to-know science, have-cause-to-know science, and personal-curiosity science (Aikenhead, 1994a; Eijkelhof & Kapteijn, 2000; Eijkelhof & Kortland, 1987).

What wish-they-knew science seems relevant to workers in science-related occupations or to public consumers making sense out of science-related information? The research reviewed above unequivocally pointed to the need to learn scientific knowledge *as required*. A key curriculum policy can therefore be affirmed: A central goal of a humanistic science curriculum should be to teach students *how to learn* Western science content as required by the contexts students find themselves in, that is, preparation for lifelong learning (Hurd, 1989b; Resnick, 1987; Roth & Désautels, 2004; Symington & Tytler, 2004). To prepare students for citizenship or science-related occupations, it would not seem to matter what Western science content is placed in the curriculum, as long as it enhances students' capability *to learn how to learn* science content within a relevant context. The selection criteria suggested by the research on relevance (reviewed in the seven categories discussed earlier) allow us to achieve the goal "to learn how to learn science content" equally well as the pipeline criterion "prerequisite coherence with first-year university courses."

Curriculum policy based on learning how to learn (lifelong learning) will produce a much different science curriculum document than will a policy based on screening students through pre-university course content. For instance, in Canada, the policy of learning how to learn is suggested by the STSE (science-technology-society-environment) approach to school science found in the Pan-Canadian Science Framework (Council of Ministers of Education of Canada, 1997).

The relevant content and contexts suggested by the seven categories are important for motivating students. However, the content and contexts may turn out to be superficially relevant for many students. A deeper meaning to relevance must be conceptualized in a curriculum policy for humanistic school science: cultural relevance.

CULTURAL RELEVANCE

As already mentioned, ideologies inherent in any science curriculum can be categorized in terms of two presuppositions of school science (Aikenhead, 2000b; Pillay, 1996; Rudolph, 2003; Weinstein, 1998): first, the enculturation of students into their *local, national, and global communities*, communities increasingly influenced by advances in science and technology; or, alternatively, the enculturation of students into the *disciplines of science*. These presuppositions represent two fundamentally different axiomatic views of relevance.

From a student's point of view, relevance concerns the degree to which curriculum content and classroom experiences speak to a student's cultural self-identity (Allen & Crawley, 1998; Brickhouse, 2001; Brickhouse et al., 2000; Brickhouse & Potter, 2001; Brown, Reveles, & Kelly, 2004; Cajete, 1999; Carlone, 2004; Chang & Rosiek, 2003; Eijkelhof, 1990; Hammond, 2001; Kozoll & Osborne, 2004; Lee & Fradd, 1998; Nicol, Tsai, & Gaskell, 2004; Stairs, 1993/94). Self-identity is a complex domain defined in a practical way by Cajete (2000a) as who we are, where we have been, where we are going, and who we want to become. In their review of self-identity and scientific literacy, Brown et al. (2004) described it theoretically by drawing upon various perspectives in psychology, sociology, anthropology, and education (Chapter 7).

In an unusually rich, in-depth, longitudinal study into student self-identities in science, Reiss (2000) examined 563 science lessons over 5 years as 22 targeted students worked their way through secondary school science in the United Kingdom. Doing science was generally seen as getting marks on the examination. "Beating the examiner" was their version of playing Fatima's rules. By interviewing students and their parents together at home (225 times in total), Reiss illuminated the cultural relevance that school science held for these students. Two unavoidable conclusions surfaced: Science education plays a meager to insignificant role in most of the students' personal lives; and school science will engage students in meaningful learning only to the extent to which the science curriculum has personal value and worth for students, that is, when it contributes to students' cultural capital (Bourdieu & Passeron, 1977; Calabrese Barton, 2004) and enriches or strengthens their self-identities (Allen & Crawley, 1998; Brickhouse, 2001; etc., cited just above). This conclusion is further validated by research into successful humanistic exhibits at science centers that personalized content, evoked emotions, stimulated dialogues or debates, and caused reflective thinking to occur (Pedretti, 2004)—in short, an experience that engaged students' self-identities.

In Mayoh and Knutton's (1997) research into using everyday experiences in science lessons, the researchers implicitly embraced the presupposition "enculturation of students into scientific disciplines," in which students' everyday experiences were deemed relevant to the extent to which those experiences motivated students to think and believe like a scientist. The 1990s' "relevance-in-science movement" (Campbell & Lubben, 2000, p. 240) similarly advanced the implicit goal to equip all students with a scientific worldview, even those students whose worldviews were incongruent with the science curriculum (Aikenhead, 1996). From the viewpoint of these students, the curriculum's goal was not enculturation into scientific disciplines but rather *cultural assimilation* (Chapter 2). As mentioned before, most students avoid this assimilation by playing Fatima's rules (Aikenhead, 2000b; Costa, 1997b; Larson, 1995).

When everyday experiences are brought into a science lesson and when students associate those experiences with their teacher's goal of cultural assimilation into scientific disciplines (not enculturation into the local, national, or global communities), most students will find those lessons irrelevant because they lack cultural value or worth (Chapter 7). This conclusion explains research referred to earlier in this chapter in which researchers reported that students' views on relevancy of everyday experiences did not match their teachers' views on relevancy (Campbell et al., 1994; Gagné, 2003).

Drawing upon culture-based research into the worldviews of a class of grade 9 students in a suburban U.S. community, Cobern and Aikenhead (1998) identified a student (Howard) who felt comfortable with the traditional school science curriculum because it harmonized with his worldview of nature. Students like Howard have future career paths enhanced by canonical science content (Lyons, 2003). This small minority of students desire to be enculturated into scientific disciplines because they see themselves possibly as future scientists or engineers. Their cultural relevance is preprofessional training that nurtures a self-identity aligned with the culture of Western science (Brown et al., 2004; Hawkins & Pea, 1987).

Cobern and Aikenhead (1998) and Costa (1995) identified many more students whose esthetic or religious worldviews of nature, for instance, were at odds with the worldview endemic to traditional school science, and whose cultural self-identities were not enhanced by a traditional science curriculum because science seemed like a foreign culture to them. Hughes's (2000) and Carlone's (2004) research, for example, shows how girls can feel culturally alienated even in some humanistic science courses because the culture of Western science is predominantly masculine, at odds with their cultural self-identities.

The foreign nature of Western science is particularly evident for most students raised in non-Western communities whose worldviews often conflict in various ways with Western science (Chapter 7). Some Aboriginal students, for instance, perceive Western science and technology as instruments used by Euro-Americans to colonize their community (Battiste, 2000; McKinley, 2003). It is common sense that the Western science taught in schools seems like a foreign culture to so many non-Western students, but Western science also seems like a foreign culture to many Western students (Aikenhead, 1997; Costa, 1995; Ogawa, 1995).

To conclude, the most fundamental question for cultural relevance is not so much "Relevant to whom?" "Relevant to what?" or "Who decides?" but rather "Relevant to which enculturation process?"—enculturation into students' local, national, and global communities (one facet of a humanistic perspective in school science advanced by this book), or enculturation into a scientific discipline (the pipeline's status quo). Cultural relevance precipitates a long-standing policy dilemma for science educators.

Some scholars have argued that the two positions are mutually exclusive in science classrooms (e.g., Hughes, 2000). Attempts to resolve this dilemma by integrating the two positions in a cross-cultural way are discussed in Chapter 7.

The research literature clearly suggests that any approach to school science, humanistic or purely scientific, dedicated to the enculturation of all students into scientific ways of thinking and believing will constantly be challenged and undermined by Fatima's rules.

PROCESSES FOR FORMULATING CURRICULUM POLICY

Throughout this book, educationally driven research agendas have conflicted with politically driven realities. When we examine the research into the processes by which people have formulated science curriculum policy, not surprisingly, politically driven realities intensify because the research tends to provoke politically sensitive questions such as, Who has the sociopolitical power to influence policy? and How do they assert and maintain that power?

The paucity of science education research in the domain of policy formulation (Hart, 2004; Knain, 2001; Kortland, 2001; Roberts, 1988) may speak to the unease felt by research participants when political events come under public scrutiny (e.g., government bureaucrats' role in policy formation). Such research exposes the natural tension between, on the one hand, maintaining a preprofessional training pipeline as well as maintaining a teacher's professional identity as a curricular technician implementing government policy and, on the other hand, innovating a humanistic perspective.

Curriculum policy is established in a number of different ways, from the "top-down" central control by government bureaucrats to the "grassroots" populist control by stakeholders (Hart & Robottom, 1990). The ultimate expression of a top-down policy formulation occurred in the United Kingdom when a national government leader publicly denounced, and therefore crushed, a humanistic perspective in the national science curriculum (Solomon, 2003b). Many curriculum policies, however, develop by way of collaboration and they attain some balance between top-down and grass-roots extremes. The balance depends on the cultural setting; some are top-down oriented (e.g., in Japan; Ogawa, 1997); some are more grass-roots oriented (e.g., in North America).

Historical events summarized in Chapter 2 revealed the cultural context in which the first formal Anglo science curriculum policy emerged in 1867 in the United Kingdom—a context characterized by the cultural values, conventions, expectations, and ideologies that determined at that time what school science would be. Because context is paramount for policy inquiry, researchers have often employed qualitative methods such as case studies or vignettes to interpret and understand processes that led to a humanistic science curriculum policy. This was certainly the case for research into power conflicts that arose during curriculum policy development, reported by Aikenhead (2002b), Barrett and Pedretti (2004), Blades (1997), Fensham (1993, 1998), Gaskell (1989, 2003), Hart (2002), Roberts (1988), and Solomon (2003b). Each study revealed a variety of power dynamics adopted by various groups of stakeholders. When people decide what knowledge is of most worth, they usually negotiate, using both rational criteria and political persuasion in an attempt to limit or enhance the influences of other stakeholders. Each educational jurisdiction has its own story to tell about how curriculum policy was formulated.

Two research studies are mentioned here to illustrate this type of research. Gaskell (1989) in British Columbia, Canada, interviewed a variety of stakeholders and showed how science teachers' allegiances to different professional organizations and to their own professional self-identities undermined an emerging humanistic science curriculum policy (Rowell & Gaskell, 1987). In his 1997 book *Procedures of Power & Curriculum Change* (a research study into the temporary defeat of a humanistic science curriculum policy in Alberta, Canada), Blades allegorically described the intense clashes between advocates of a humanistic science curriculum and newly aligned interest groups who organized a network of relationships (actor-networks; Carlone, 2003; Foucault, 1980; Gaskell & Hepburn, 1998) to serve their own self-interests, and who enacted "rigor" as a power ploy in their discourse. Blades discovered that, in the midst of developing a new STS curriculum for the high school, one very powerful stakeholder group reformulated its alliances along different lines, thereby reversing its policy position. Treachery thy name is government bureaucrat! Both research studies (in British Columbia and Alberta) provide answers to the questions, Who has the power to decide? and How do they assert and maintain that power?

Although each case study and vignette in the research literature was unique, they reached the same conclusion (with a few exceptions): Local university science professors have a self-interest in maintaining their discipline and will boldly crush humanistic initiatives in school science policy (Aikenhead, 2002b; Blades, 1997; Fensham, 1992, 1993, 1998; Fensham & Corrigan, 1994; Gaskell, 1989; Hart, 2001, 2002; Panwar & Hoddinott, 1995). Gaskell (2003) called this "the tyranny of the few" (p. 140). If local science professors become marginalized and lose their power to control policy decisions, they tend to realign their actor-networks into international alliances to defeat a local humanistic curriculum policy (Rafea, 1999). Sometimes they resort to blackmail (Aikenhead, 2002b) or raw political power to control bureaucrats (Hart, 2001).

Science curriculum policy is normally formulated more smoothly through consultation with different stakeholders (Orpwood, 1985), for instance, government officials, the scientific community, science teachers, university science educators, parents, plus others representing groups and institutions in society. Government departments of education generally rely on the advice of curriculum committees variously comprised of some of these stakeholders. Because government committee meetings are almost always held out of view of an inquisitive researcher, their confidentiality has prevented research into the early stages of formulating government policy (De Vos & Reiding, 1999; Roberts, 1988). An interesting exception is Boydston's (1999) in-depth critical investigation of a large urban school district's attempt at reforming its traditional school curriculum. His insider status afforded him interviews and insights not enjoyed by an outside researcher. He discovered a type of naïve self-deception by some stakeholders who did not realize that they were actually policy makers and not simply policy implementers. Their naïvety made them vulnerable to manipulation by other stakeholders.

Curriculum policy research has taken the form of research and development (R&D) studies that produce classroom materials (e.g., textbooks and modules) as a means to influence or articulate a humanistic curriculum policy (Chapter 4). Researchers have collaborated with departments of education, teachers, and students, as well as with experts who furnished functional and have-cause-to-know science for the curriculum (Aikenhead, 1994a; Eijkelhof & Kapteijn, 2000; Eijkelhof & Lijnse, 1988; Kortland, 2001).

Along similar lines, consultation with Aboriginal elders is a key feature of policy development to achieve cultural relevance, sensitivity, validation, and support in an Aboriginal community (Aikenhead, 2002a; ANKN, 2004a; Inuit Subject Advisory Committee, 1996; Kawagley, Norris-Tull, & Norris-Tull, 1998; McKinley, 1996; Riggs, 2005; Stephens, 2003; Sutherland & Tays, 2004). Elders and community members are seen as authoritative stakeholders in any deliberation over curriculum policy.

Another type of collaboration is grounded in classroom action research (Hodson, Bencze, Elshof, Pedretti, & Nyhof-Young, 2002). This research methodology has produced innovative curriculum policies along with support materials from, for instance, Solomon et al. (1992) in the United Kingdom, Keiny (1993) in Israel, and Pedretti and Hodson (1995) in Canada. Culture studies in science education have combined action research with anthropological research to produce culturally relevant policies and practices for students conventionally marginalized in school science (research reviewed by Hammond & Brandt, 2004). Policy negotiations have occurred through action research in urban science education (Calabrese Barton,

2004), in multicultural science education (Hammond, 2001), and in working-class communities (Gilbert, 2002). On the one hand, action research can holistically address curriculum policy development along with the concomitant teacher professional development (Hodson et al., 2002), a topic for Chapters 4 and 5. But, on the other hand, action research involves only a tiny proportion of excellent teachers, and its products do not transfer easily to other teachers (Jenkins, 2004; Solomon, 1999b).

Some researchers have used the Delphi research method to inform humanistic curriculum policy, for instance (as described earlier), the research by Eijkelhof (1990), Häussler and Hoffmann (2000), and Osborne, Collins, et al. (2003). Their experts were able to reach a consensus on the relevant contexts and associated knowledge for an educationally sound, humanistic science curriculum policy. The lack of face-to-face collaboration in Delphi research, however, raises questions concerning the clarity of the statements on which they seem to agree.

A different ethnographic research method can also give voice to stakeholders not normally invited to policy debates. Symington and Tytler's (2004) study, mentioned earlier, tapped into the cultural capital and political power held by their sample of high-profile *Who's Who* participants in their study.

The most elaborate, theory-based, consultative methodology, however, is deliberative inquiry. It combines top-down and grass-roots approaches. Deliberative inquiry is a structured and informed dialogic conversation among stakeholders who, face to face, help government officials reach a decision on curriculum policy by discussing and reexamining their own priorities (i.e., values) along with their reading of relevant research (Orpwood, 1985). Because science teachers will be central to implementing a humanistic science curriculum (Roberts, 1988) and because curriculum evaluation research consistently shows that the teacher has more influence on student outcomes than does the government's choice of curriculum (Welch, 1979, 1995), the science teacher is a key stakeholder and usually holds a central role during deliberative inquiry meetings. The process of deliberation encompasses both educational and political dimensions to formulating curriculum policy.

The Science Council of Canada (SCC) used deliberative inquiry to produce a national science curriculum policy that embraced a humanistic perspective (Aikenhead, 2000c; Orpwood, 1985). The SCC study ensured that significant problems in science education were identified, that appropriate evidence was collected, and that the problems and evidence were considered by diverse stakeholders attending one of the many 2-day deliberative conferences held across Canada. Stakeholders included high school students (science-proficient and science-shy), teachers (elementary and secondary), parents, elected school officials, the scientific community (academic and industrial), university science educators, and representatives from the business and labor communities. The students' contributions were pivotal to recommendations related to student assessment (SCC, 1984). As Schwab (1978) predicted, "Deliberation is complex and arduous. . . . [It] must choose, not the *right* alternative, for there is no such thing, but the *best* one" (pp. 318–319, emphasis in original). The "best" science curriculum policy for Canada was published as *Science for Every Citizen* (SCC, 1984). Inspired by the success of this deliberative inquiry, the Canadian province of Saskatchewan replicated some of the SCC study during the renewal of its science curriculum and yielded a strong teacher consensus on a humanistic perspective (Hart, 1989).

A very different method of policy formulation, illustrated by the AAAS's (1989) *Project 2061* in the United States, utilizes consultation with stakeholders on a grand yet narrow scale. After conducting a complex series of inclusive national surveys and committee meetings, a "consensus panel of leading scientists" (Walberg, 1991, p. 57) determined the content of *Project 2061*, content critiqued as conveying an outmoded positivist and noncontemporary view of science (Bingle & Gaskell, 1994; Fourez, 1989; Hurd, 2002); as ignoring student relevancy (Settlage & Meadows, 2002); as overlooking students' worldviews, self-identities, and cultural diversities (Calabrese Barton & Osborne, 2001; Lee, 1997; Lemke, 2001; Lynch, 2001); and as not clarifying its own political and cultural agendas (Calabrese Barton & Osborne, 2001; Kyle, 1996). Thus the final say in the AAAS's curriculum policy was greatly influenced by people who generally espouse the conventional wish-they-knew science. This exclusivity, plus the lack of published research on the consultation process itself, suggests that the national agency may have prioritized political realism over educational soundness and thereby repeated its predecessors' 1867 policy decision (Chapter 2). A humanistic perspective loses significance in the predominant wish-they-knew science of *Project 2061* (AAAS, 1989) and *Standards* (NRC, 1996) in spite of humanistic rhetoric posited in their policy statements.

CONCLUSION

An education jurisdiction's decision on what knowledge is of most worth arises from a number of historical, social, cultural, and political forces. This chapter's synthesis of research into these forces did not produce a simple image of curriculum policy development but rather an interactively dynamic complex of individuals who unconsciously or purposefully represented a variety of institutional stakeholders, including students, teachers, other employees of school systems, parents, community employers, university science/engineering professors and registrar offices, university science educators/researchers, professional scientific organizations and institutes, teachers' professional organizations, spokespersons for national businesses and industries, and government bureaucrats. Some are close to science classrooms. Some represent social structures in which classrooms are expected to operate. Others are far removed from current classroom life. Understandably, each has different self-interests. The ideology to exert the most power over curriculum policy will be the ideology most forcefully asserted in meetings of curriculum decision making. The simple ideology of preprofessional training of scientists and engineers seems to have political advantages over more complex ideologies that inspire humanistic approaches to school science.

"Policy changes that ignore the interconnections of ideologies and [social and cultural] structures have little chance of acceptance" (Krasilchik, 1995, p. 423). Considering all the research studies into policy formulation reviewed here, the process of deliberative inquiry holds greatest potential for advancing an educational rationale for a humanistic perspective in the science curriculum because it provides a political forum for negotiation among various stakeholders. "This requires bringing to the surface the tacit social aims and assumptions that are constantly in play in the development of the school science curriculum as well as carefully considering the social consequences, intended or not, of the curriculum produced" (Rudolph, 2003, p. 76).

Although these considerations will likely enhance the quality of the resultant science curriculum policy, other important areas of research are pertinent to successful deliberative inquiry. Some of these are classroom materials (Chapter 4), teacher orientation (Chapter 5), and student learning (Chapter 6).

4

Classroom Materials

Classroom teaching materials, particularly textbooks and modules, seem to dictate the taught curriculum for many teachers (Chiang-Soong & Yager, 1993; Lijnse, 1995; Osborne, Duschl, et al., 2003). Textbooks and modules, therefore, have the potential to operationally define humanistic science teaching and guide teachers who are implementing a humanistic perspective in their classroom. In the absence of such guidance, teachers feel unsupported (Bartholomew et al., 2004).

RESEARCH ANALYZING CLASSROOM MATERIALS

Roberts (1982) systematically analyzed a number of North American science policy statements woven into textbooks during the 20th century to determine the subtle messages they conveyed to students about goals for studying science. His research inductively developed seven different messages he called "curriculum emphases," a concept used widely in science education today. Fensham (2000b) added three more emphases to Roberts's list, based on similar research into recent innovative textbooks. Although the most popular curriculum emphasis for secondary schools was "solid foundations" (preprofessional training), humanistic emphases did appear: one in Roberts's study ("science, technology, and decisions") and in all three of Fensham's additions ("science in application," "science as nurturing," and "science through technology"). Textbooks over the years generally did not ignore a humanistic perspective, but they gave it very meager attention until recently.

As the STS school science movement evolved in the 1980s, researchers investigated the degree to which STS content appeared in contemporary science textbooks (e.g. Chiang-Soong & Yager, 1993). Bybee (1993) reviewed this literature and concluded that little STS content could be found, hence indicating a low value placed on it by textbook authors and commercial publishers. More important, however, his review, along with Monk and Osborne's (1997) research in the United Kingdom, noted that even when STS textbooks and modules were generally made available to teachers, the materials were not implemented to any extent. These findings bring us to a key conclusion: The availability of humanistic science materials is a necessary but insufficient condition for a humanistic perspective to be found in the taught curriculum (Cross & Price, 1996; Eijkelhof & Kortland, 1987), a topic explored in Chapter 5.

Researchers have also analyzed traditional science textbooks to investigate the images they conveyed *about* science and scientists (Cross & Price, 1999; Gaskell, 1992; Kelly et al., 1993; Knain, 2001; Milne, 1998). The results showed a clear picture of idealized, heroic, highly rational, individual scientists who reveal truth by applying the scientific method—a picture that mythically equates scientific knowledge of nature with nature itself. Most high school science textbooks attempt to indoctrinate the reader into an ideology of positivistic realism endemic to the traditional science curriculum (Anderson & Kilbourn, 1983; Östman, 1996). On the other hand, similar research into a humanistic chemistry textbook, *ChemCom*, concluded that although it contained vestiges of this traditional ideology, students were not necessarily being indoctrinated if a teacher took a reasonably critical stance in the classroom (Carlsen, Kelly, & Cunningham, 1994).

By analyzing the humanistic textbook *Science and Sustainability* (SEPUP, 2003), Ratcliffe (2004) documented the difficulty 22% of the teachers had when they tried to integrate its canonical science content with the social issues related to sustainability. Based on teachers' written analyses of various aspects of the textbook, Ratcliffe concluded, "There is a demanding and, to some extent, unresolved tension . . . in devising and teaching a course which adequately addresses both citizenship and the understanding of science concepts" (p. 9). Her data revealed that teachers' positive reaction to the textbook depended on the degree to which they supported humanistic outcomes in school science, an issue examined in Chapter 5.

Classroom materials similar to *ChemCom* and *Science and Sustainability* that follow an issues-based (humanistic) approach can be undermined by political manipulation. Hughes (2000) discovered that science educators who embraced a pipeline ideology were subtle but effective in devaluing and marginalizing humanistic content by the way canonical science content was emphasized through the language and organizational structures of classroom materials. This was also a general problem identified with materials developed during the 1990s' relevance-in-science movement (Campbell & Lubben, 2000).

Emphasis and marginalization are also issues addressed in cross-cultural classroom materials. Ninnes (2000) investigated how indigenous science was represented in Australian and Canadian textbooks specifically written to include Aboriginal science. He discovered that the authors tended, first, to stereotype Aboriginal peoples (i.e., inadvertent racism); second, to prescribe (by the authors' own authority) what is authentically traditional in Aboriginal cultures (i.e., a form of colonization); third, to use only the past verb tense when describing Aboriginal science, thereby undermining its current relevancy; and fourth, to subjugate Aboriginal epistemologies to Western scientific epistemology (i.e., a form of colonization). All of these features cause classroom materials to be instruments of "cultural imperialism," that is, they attempt to assimilate (colonize) Aboriginal students into Western scientific ways of knowing—an antihumanistic ideology to be sure. This topic is discussed in greater detail in Chapter 7.

Research into humanistic science materials exemplifies *formative* and *summative* assessment (Black, 1998). Broadly speaking, formative assessment produces feedback from teachers and students to textbook writers, identifying problems to be resolved to improve the materials. Summative assessment, on the other hand, usually produces data on student learning and teacher satisfaction for the purpose of evaluating the overall effectiveness of the materials. Summative assessment data can be useful to potential teachers as well as to policy makers who can offer political support for the innovation. This chapter reviews studies in formative assessment, leaving summative assessment to Chapter 6.

RESEARCH INTO THE DEVELOPMENT
OF CLASSROOM MATERIALS

Systematic investigations into the development of humanistic science textbooks or modules have been rare because the financial resources are usually spent on the production of the materials themselves and on the professional development programs for prospective teachers. Even Harvard Project Physics (Chapter 2) did not systematically research the *development* of its textbook, and as a consequence the writers did not revise it sufficiently to meet the needs of many teachers and students (Cheek, 2000; Welch, 1979, 1995).

Research projects reviewed in this chapter illustrate various degrees of collaboration between writers/developers of classroom materials (i.e., those who have a vision of the intended humanistic science curriculum) on the one hand and, on the other, teachers and students who are expected to engage with those materials (i.e., the translated and learned curricula, respectively). Collaborative research methodologies have included research and development (R&D), developmental research, and action research.

Research and Development

In the natural sciences, R&D combines scientific inquiry and engineering design in a context bounded by everyday exigencies (Ziman, 1984). In the social science domain of education, R&D produces data that are fed directly into improving the product of the study or improving how people use it. This type of research differs from the typical science education research reported in the literature where data are collected to inform a theoretical model or interpreted to convey a participant's lived experience, for instance. Because R&D studies are dedicated to improving a product or technique, they are rarely written up for publication in the research literature.

The most substantial R&D study to publish its formative assessment took place in The Netherlands from 1972 to 1986. The PLON project developed many humanistic physics modules for lower and upper secondary school (Eijkelhof & Kortland, 1987; Eijkelhof & Lijnse, 1988). The modules were expected to motivate students to learn canonical physics content by placing that content in relevant contexts (e.g., the physics of sound was taught in a module called "Music"). The modules also aimed to improve students' capacity to interpret media messages, to make consumer choices, to follow new developments reported in the media, and to engage in public decision making, all related to physics. The researchers used questionnaire and interview data from students and teachers, as well as field notes of teacher meetings and occasional classroom observations.

At least three major results emerged from this research program. The first concerns the difficulty the PLON developers encountered finding real-life contexts that involved only canonical physics content, owing to the fact that science-related events in the real world are largely interdisciplinary (e.g., music touches significantly on physiology and esthetics). This situation reflects the historical evolution of 19th-century scientific disciplines into 20th-century collectivized science (Chapter 2). The PLON project was constricted by a 19th-century national curriculum policy that required separate disciplines, rather than allowing for the integration endemic to collectivized science. The second major result of the PLON research program concerns student learning: Students' meaningful learning of canonical physics did not improve significantly in PLON units, nor did it suffer (Eijkelhof, 1990). This illustrates the conclusion that most students experience inherent difficulty achieving meaningful learning in school science (Chapter 3). The third major result of PLON's research program was its recommendation that future R&D projects conduct in-depth investigations into problems related to humanistic science materials *in general*, while research is being undertaken to develop specific classroom materials. PLON's R&D studies have provided sound evidence and valuable advice to science educators worldwide.

Other R&D studies have involved students more directly in the development of classroom materials (Aikenhead, 1994a, 2000c; Solomon, 1983). For instance, in my R&D production of an integrated STS science textbook, *Logical Reasoning in Science & Technology*, I collaborated extensively with grade 10 students by teaching the first version myself, a version to which students actively contributed material (Aikenhead, 1994a). Next, I observed three teachers daily as they taught a draft of the student text, supported only by a draft teacher guide. This resulted in refinements to both documents. These classroom materials developed in situ, as daily research on student preconceptions of canonical concepts, for instance, immediately led to changes in the text, as suggested by the practical actions of students and teachers. The next revision was piloted by a diverse group of 30 teachers whose written and focus-group feedback fine-tuned the materials for commercial publication. This R&D project established that students can contribute significantly to a textbook's content, structure, and language, and that most students respond eagerly to classroom material produced this way (Aikenhead, 1994a). But unlike in the PLON project, resources were available only for formative assessment research.

Other published research into the formative assessment of materials occurred for a humanistic biotechnology curriculum in Spanish high schools (Sáez, Niño, Villamañan, & Padilla, 1990), for a New York state STS project (Cheek, 1992), and for three issues-based humanistic textbooks implemented throughout the United States by the Science Education for Public Understanding Program (SEPUP) (Thier & Daviss, 2001; Wilson, 2004).

Read (1998) documented the valuable collaboration between himself and Aboriginal elders and community members in Australia during his R&D study that produced *The Kormilda Science Project: An Indigenous Perspective on the Earth Sciences* (Read, 2002), a rare example of a textbook that respectfully integrates indigenous science with Western science. Cross-cultural science projects have typically produced guidelines describing indigenous knowledge (Chapters 3 and 7), but they have not conducted research into how science teachers follow those guidelines to develop materials subsequently used in their classrooms.

Developmental Research

The third recommendation from the PLON's R&D studies led to a new methodology, *developmental research*: "a cyclic process of reflection on content and teaching/learning process, small-scale curriculum development and teacher preparation, and classroom research of the interaction of teaching and learning processes" (Kortland, 2001, p. 10). These are research studies that produced PLON-like modules, but their validity rests on the study's transferability to other teaching contexts (e.g., the improvement of science teaching in other settings, and the development of didactic theories) rather than resting only on the quality of the specific materials produced (Eijkelhof, 1990; Kortland, 2001; Lijnse, 1995). For instance, Eijkelhof (1990) demonstrated how a Delphi study could generate curriculum content (in this case, have-cause-to-know science; Chapter 3) and some concomitant authentic contexts for teaching this content. His study resulted in the development of a module called "Ionizing Radiation," which integrated the topics of risk and safety assessment with physics.

Kortland's (2001) work also illustrates developmental research. He studied the phenomenon of student decision making (Chapter 6) while he developed an ecology module on waste management. The production of quality classroom materials is only one key outcome to developmental research.

Developmental research laid the foundations for producing a grade 10 humanistic science curriculum in The Netherlands between 1996 and 2003, *Algeme Natuurwetenshappen*, often referred to as "the public understanding of science" curriculum (De Vos & Reiding, 1999; Eijkelhof & Kapteijn, 2000). Interestingly, the final responsibility for polishing these classroom materials was turned over to commercial publishers, who do not publish their internal research.

Action Research

Action research is small-scale classroom R&D usually initiated by teachers to find solutions to their practical problems or to initiate curriculum change in their own classroom (Hodson et al., 2002). By combining the production of educationally sound knowledge with politicization (Hodson, 1994; Keiny, 1993), action researchers have produced practical humanistic science materials and have influenced, to some measure, their country's science curriculum (Aikenhead, 2000c, 2002a; Hammond & Brandt, 2004; Hodson et al., 2002; McFadden, 1996; Ogborn, 2002; Pedretti & Hodson, 1995; Rye & Rubba, 2000; Yager, 1996). Following the completion of an action research study, the original participants can become actively involved in developing new materials with other teachers. However, as mentioned in Chapter 3, a general limitation to action research stems from the fact that only a small proportion of teachers are involved, a situation that does not adequately meet the needs of most teachers (Solomon, 1999b). And unfortunately many classroom materials produced by action research are quite idiosyncratic and this inhibits the materials' transferability to other teachers' classrooms (Jenkins, 2004).

There are notable success stories in action research, nevertheless. The Atlantic Science Curriculum Project worked 15 years to publish a Canadian humanistic textbook trilogy *SciencePlus* (McFadden, 1991, 1996) and then an American textbook trilogy *SciencePlus Technology & Society* (McFadden & Yager, 1997). Other successes include the Science and Technology in Action in Ghana Project (Anamuah-Mensah, 1999); the Science, Technology, Environment in Modern Society (STEMS) project in Israel (Dori & Tal, 2000; Tal, Dori, Keiny, & Zoller, 2001); and the Science Through Applications Project in South Africa (Gray, 1999). The Israeli and South African projects produced a number of modules by teams of teachers supported by parents and the communities' industries, over a 1-year and 3-year period, respectively. The Israeli and South African projects' formative assessment targeted teachers' professional growth but not the classroom materials themselves. By involving people from the community in the development of materials, the Israeli and South African projects benefited from the knowledge found in the local culture (e.g., indigenous science, functional science, and science-as-culture). This knowledge was incorporated into the classroom materials.

A more modest action research project in Canada, *Rekindling Traditions* (Aikenhead, 2000a), investigated ways to engage Aboriginal communities in collaborating with a science teacher and students to develop local classroom materials that integrated Aboriginal science with Western science (Aikenhead, 2002a). This project explicitly avoided the problems identified in Ninnes's (2000) research (discussed above). The resultant six teaching units illustrate a cross-cultural orientation to a humanistic perspective in the science curriculum, from which the traditional science curriculum is viewed as a foreign culture to be appropriated by students for the benefit of their Aboriginal cultures (a topic elaborated upon in Chapter 7). Gagné (2004) analyzed some of the *Rekindling Traditions* units through the lens of students' cultural self-identities. She concluded that the teaching materials honored "personal and group identities by trying to enculturate students into their own community rather than trying to assimilate them into the community of Western science" (p. 11)—in other words, cultural relevance (Chapter 3). People in each indigenous community helped decide what should be taught and they helped teach it (Aikenhead, 2002a), illustrating a community-based method for developing classroom materials (Maddock, 1981; May, 1999).

The *Rekindling Traditions* project inspired others. A Canadian Cree community conducted action research to develop three of their own culturally responsive units (Sutherland & Tays, 2004), and 60 Māori preservice education students in Aotearoa New Zealand developed resource materials that sensitively bridged the students' Mäori cultural identity and the culture of science (Barker, 2004).

Similarly, Malcolm and Keogh (2004) produced a unit built on the interests of students and their families in a rural Zulu community in South Africa. Indigenous science (in this case Zulu traditional knowledge of clouds, plants, and animals) was integrated with Western science knowledge. Classroom materials told stories about rain, which required students to distinguish between *predicting* rain and *making* rain, an epistemological issue that arose when the class compared Zulu thinking with Western thinking.

In Nepal where the traditional science curriculum was also seen to clash with students' cultural self-identities, Bajracharya and Brouwer (1997) demonstrated how collaborative action research by teachers, students, administrators, and community members can develop classroom materials called "narratives" that promote a respectful integration of indigenous science with the curriculum's Western science content.

CONCLUSION

Humanistic science materials have been produced in sufficient numbers to guide future projects, yet there continues to be a paucity of materials for classroom use.

Research into humanistic teaching materials has dealt mainly with the production of innovative materials. The research did not investigate the materials' influence on teachers, for example, by asking, To what extent did the materials influence the taught curriculum enacted by the teachers? To what extent did teachers modify the materials to conform with the teachers' personal orientation to a humanistic science curriculum? or To what extent did the materials inspire teachers to modify their practical knowledge of teaching and their classroom practice? These and other crucial questions are explored in Chapter 5.

5

Teacher Orientation

Teachers construct their own meaning of any intended curriculum as they negotiate an orientation toward it and decide what to implement, if anything, in their classroom. Over the years, researchers have studied teachers' rejection, acceptance, and idiosyncratic modulation of an intended humanistic science curriculum. In the post-Sputnik era of the 1960s, simplistic assumptions about "teacher-proof" curricula predominated (Welch, 1979). Our assumptions about teachers' influence on the taught curriculum have evolved into more complex researchable frameworks that describe how teachers make decisions and take action. My synthesis of research in this chapter stresses recent studies.

The historical research in science education (Chapter 2) predicts that a small proportion of science teachers will always be predisposed, in varying degrees, to an ideology supportive of a humanistic perspective for school science. Thus there will always be a few science teachers who teach from a humanistic point of view (humanistic science teachers) and who gladly participate in any research study that promises to enhance their humanistic orientation. History similarly predicts there will be a nucleus of teachers committed to the pipeline ideology that promotes preprofessional training, mental training, and screening students for university entrance. These teachers (pipeline enthusiasts) will resist and perhaps actively challenge any humanistic innovation in school science (Blades, 1997; Carlone, 2003; Fensham, 1992; Rowell & Gaskell, 1987). There exists a third group of science teachers who can be persuaded to move toward either ideology for a variety of reasons (middle-of-the-road teachers). Hence teacher participants in any research study may be identified by these three categories: humanistic, pipeline enthusiast, and middle-of-the-road (McGrath & Watts, 1996; Mitchener & Anderson, 1989); 15%, 11%, and 74%, respectively, for Dutch chemistry teachers (van Driel, Bulte, & Verloop, 2005).

Conclusions from research will be affected by the presence and proportion of teachers from these three groups. For instance, in one of the most insightful inservice programs for a humanistic science curriculum

(Leblanc, 1989), its leaders ensured that a high proportion of participants came from humanistic and middle-of-the-road teachers, and the leaders judiciously selected a small number of high-profile teachers from the pipe-line-enthusiast group. After 3 years of periodic, intensive inservice sessions, supported by university research scientists and enriched by classroom trials of materials by participants and then followed by in-depth group reflection, the province of Nova Scotia, Canada, formally implemented an STS science curriculum supported by all the inservice teachers, including the pipeline enthusiasts. No follow-up study into the sustainability of this humanistic curriculum was reported, however.

Research into science teachers' orientations to a humanistic approach is synthesized here in terms of challenges to teachers' acceptance of a humanistic curriculum, teachers' decisions *not* to implement it, teachers' success at implementing it, components of a teacher's orientation, preservice experiences in teacher education, and the power and politics of school culture.

CHALLENGES TO TEACHERS' ACCEPTANCE

Normally, science teachers have been attracted to, and uniformly socialized into, specific scientific disciplines in university programs where teachers become certified loyal gatekeepers and spokespersons for science; and in return they enjoy high professional status and a self-identity associated with the scientific community (Bartholomew et al., 2004; Cross, 1997a; Cross & Ormiston-Smith, 1996; Cross & Price, 2002; Gallagher, 1998; Gaskell, 1992, 2003; Hughes, 2000; Roberts, 1988; Venville et al., 2002). As a consequence, teachers tend to favor abstract decontextualized "pure science," and they tend to marginalize student-oriented perspectives and utilitarian aims related to everyday life. A teacher's loyalty to the academic science community and to its cultural myths is manifested in a teacher's strong orientation to a traditional science curriculum (Abd-El-Khalick & BouJaoude, 1997; Aikenhead, 1984; Allchin, 2003; Davis, 2003; Duschl, 1988; Kilian-Schrum, 1996; Milne & Taylor, 1998). This orientation engenders the transmission of established knowledge and technique (canonical science content) to persuade students of the correctness of a scientific worldview steeped in positivism and realism (Bartholomew et al., 2004; Osborne, Duschl, et al., 2003).

When STS was proposed for school science, Gaskell (1982) and Gallagher (1987) questioned the capability of science teachers steeped in empirical reductionist worldviews to engage competently with ethical, economic, and political issues in the classroom. Although research has by and large confirmed these concerns, it has afforded helpful insights into the complex world of science teaching.

Salient influences on a teacher's orientation have been documented by researchers. These influences include a teacher's values, assumptions, beliefs, ideologies, self-identities, self-images, and loyalties to traditional school science. All research unanimously and unambiguously confirms one result: *Changing any one of these salient influences toward a humanistic perspective is very difficult* for most middle-of-the-road teachers, and is usually impossible for pipeline enthusiasts (Aikenhead, 1984, 2000c; Anderson & Helms, 2001; Briscoe, 1991; Cronin-Jones, 1991; Cross, 1997a; Davis, 2003; Gallagher, 1991; Hart & Robottom, 1990; Helms, 1998; Kortland, 2001; Lantz & Kass, 1987; Lumpe, Haney, & Czerniak, 1998; McRobbie & Tobin, 1995; Mitchener & Anderson, 1989; Osborne, Duschl, et al., 2003; Roberts, 1998; Sáez & Carretero, 2002; Tobin & McRobbie, 1996; Walberg, 1991; Yerrick, Parke, & Nugent, 1997).

Taken together, this cluster of salient influences has been referred to as the culture of school science (Aikenhead, 2000b; Bianchini & Solomon, 2003; Lee, Choi & Abd-El-Khalick, 2003; Medvitz, 1996; Munby, Cunningham, & Lock, 2000; Pedersen & Totten, 2001; Solomon, 1994e, 2002; Tobin & McRobbie, 1996; Venville et al., 2002). Implementing a humanistic science curriculum is judged by teachers to be either culturally safe or unsafe (McGinnis & Simmons, 1999). "These teachers are moulded by the culture and habitus of the culturally accepted practice of science teaching—an activity in which they have engaged, often for many years. Breaking that mould is, therefore, neither straightforward nor simple" (Osborne, Duschl, et al., 2003, p. 11). The mold seems to be similar the world over, based on the evidence that science teachers in diverse countries share very similar worldview presuppositions (Ogunniyi, Jegede, Ogawa, Yandila, & Oladele, 1995) and teach in similar ways (Gallagher, 1998).

Teachers often speak about a humanistic science curriculum in terms of their comfort level. Researchers Barnett and Hodson (2001) concluded, "Knowledge that enables teachers to feel more comfortable in the classroom and to enhance their sense of self is likely to be embraced; knowledge that increases anxiety or makes teachers feel inadequate will almost certainly be resisted or rejected" (pp. 431–432).

DECISIONS NOT TO IMPLEMENT

When asked if teaching from a humanisticlike perspective is a good idea, most science teachers (about 90%) overwhelmingly endorse it (Bybee, 1993; Hart, 1989; Lee et al., 2003; Pedersen & Totten, 2001; Rubba, 1989;

van Driel et al., 2005). Yet when asked about implementing such a curriculum, teachers provide many reasons for not doing so.

Their reasons are posited here, not in any particular order of importance because the presence and priority of these reasons change from study to study. A lack of available classroom materials is often mentioned but, as stated in Chapter 4, when teaching materials do become available, teachers point to other reasons for not implementing a humanistic approach. These reasons are unfamiliarity with student-centered, transactional, or transformative teaching and assessment methods (including group work or divergent thinking); greater than normal emphasis on oral and written language, and the complexity caused by combining everyday and scientific genres; lack of confidence with integrated content and with connecting science content to students' everyday lives; fear of losing control over the class (e.g., open-ended activities and unpredictable outcomes— teachable moments); uncertainty about a teacher's role in the classroom (e.g., as a facilitator) in spite of attending inservice workshops; a reliance on a single national textbook that contains little or no humanistic content; pressure to comply with state content standards defined by the current reform movement; an unease with handling controversial issues, or even group discussions of a social or ethical nature; a perception that science is the search for truth, so ethical issues are not appropriate content; uncertainties over assessing students on "subjective" content; inadequate background knowledge and experiences (e.g., preservice teacher education programs); no opportunity to work with an experienced competent teacher or with scientists in industry; lack of school budget to support the innovation; lack of administrative or collegial support; lack of parental or community support; no clear idea what the humanistic innovation means conceptually or operationally; predictions that students will not appreciate or enjoy philosophical, historical, and policy issues in a science class (e.g., "students want to light Bunsen burners and get the right answer"); a preoccupation with preparing students for high-stakes examinations and success at university; pressure from university science departments to raise standards and cover more content in greater depth; teachers' high status as custodians of scientific knowledge and gatekeepers to higher learning; an unease over the reduced time devoted to canonical science content and to covering the traditional curriculum; believing humanistic content devalues traditional school science; identifying oneself with scientists (i.e., as a lecturer expert) rather than with educators; the fact that nonelite and low-achieving students enroll in humanistic science courses; perceiving a threat to science teachers' privileged social status within the school and community; greater need for cultural sensitivity with some humanistic topics, such as social justice, in the use of science and technology; conformity to the school's local culture; and, finally, the survival mode of beginning teachers that discourages them from taking seriously humanistic ideas developed in their teacher education courses (Aikenhead, 1984, 1994a; Bianchini, Johnston, Oram, & Cavazos, 2003; Bybee, 1993; Bybee & Bonnstetter, 1987; Cajas, 1999; Cross & Price, 1996; Driver, Newton, & Osborne, 2000; Eijkelhof, 1990, 1994; Eijkelhof & Kortland, 1987; Gallagher, 1987; Gaskell, 1992; Gray, 1999; Gray & Bryce, 2003; Hodson et al., 2002; Hughes, 2000; Levinson, 2004; Levinson & Turner, 2001; McClelland, 1998; McGinnis & Simmons, 1999; McGrath & Watts, 1996; Mitchener & Anderson, 1989; Monk & Osborne, 1997; Munby et al., 2000; Osborne, Duschl, et al., 2003; Pedersen & Totten, 2001; Pedretti, 2003; Reis & Galvão, 2004; Reiss, 1999; Rigano & Ritchie, 2003; Roberts, 1988, 1998; Schwartz & Lederman, 2002; Solomon, 1990; 2002; Tsai, 2001a; Walberg, 1991). We are faced with an inescapable conclusion: *There are daunting problems facing educators who wish to change the traditional science curriculum into a humanistic one.*

Even under circumstances supportive of teaching from a humanistic perspective, willing teachers will modulate a humanistic science textbook or other resources to conform to their goals for teaching science and to reflect their personal epistemic and sociological beliefs (Aikenhead, 1984; Barnett & Hodson, 2001; Carlone, 2003; Ryder, Hind, & Leach, 2003a). Hlady (1992) documented how science teachers consciously omitted textbook passages that offered ideas at odds with the teachers' understanding; reinterpreted passages for their students in a way that changed the textbook's meaning completely; or personalized sections of the textbook in such a way as to change its intent—for instance, they presented an anecdotal lecture to avoid the decision-making perspective intended by the textbook.

A number of research studies (from surveys to case studies) have focused on teachers' prerequisite knowledge of humanistic content as a possible determinant to their decision to teach that content (Abd-El-Khalick & Lederman, 2000; Cunningham, 1998; Cross & Price, 1996; Gallagher, 1991; Lederman, 1992; Pedersen & Totten, 2001; Rubba, 1989; Rubba & Harkness, 1993), especially when a humanistic perspective is first introduced into a country, such as in South Korea (Lee et al., 2003) or Lebanon (Abd-El-Khalick & BouJaoude, 1997). Although the results generally showed inadequate and discrepant understanding by teachers, the results also showed that an adequate understanding was not necessarily a prerequisite to a teacher's decision to implement a humanistic perspective (Abd-El-Khalick, Bell, & Lederman, 1998; Bartholomew et al., 2004; Tsai, 2001a). At one extreme, for example, some teachers gained an understanding of humanistic content only through implementing a humanistic perspective in their classrooms (Aikenhead, 2000c; Bencze & Hodson, 1999; Fensham & Corrigan, 1994; Ratcliffe, 2004; Sáez & Carretero, 2002; Tal et al., 2001;

Tsai, 2001a). These could have been middle-of-the-road teachers. At the other extreme, some teachers (pipeline enthusiasts, perhaps) maintained their traditional preconceptions (e.g., positivism or scientism) in spite of explicit instruction in humanistic content by curriculum specialists, or in spite of teaching a humanistic science course over a period of time (Gallagher, 1991; Hlady, 1992; Yerrick et al., 1997). In short, a prerequisite understanding may or may not be a necessary condition, although by itself it is certainly not a sufficient condition, to influence a teacher's decision to implement a humanistic perspective.

A complementary avenue of research looked at science teachers who possessed a contemporary understanding of humanistic content but whose taught curriculum did not convey that content to students (Abd-El-Khalick et al., 1998; Brickhouse, 1989; Lederman, 1992; McComas, Clough, & Almazora, 1998; Southerland, Gess-Newsome, & Johnston, 2003). These studies listed specific constraints felt by teachers knowledgeable in humanistic content, constraints found in the plethora of reasons (stated earlier) given by teachers for not implementing a humanistic science curriculum. Other researchers discovered more pervasive constraints, such as the school's social organization and the school culture (Carlone, 2003; Gray & Bryce, 2003; Marchetti & Camino, 2003; McGinnis & Simmons, 1999; Medvitz, 1996; Rigano & Ritchie, 2003). These constraints (discussed near the end of this chapter) can overwhelm humanistic teachers' best intentions to include a humanistic perspective in classroom practice—their taught curriculum. The research results certainly challenge the simplistic notion that teachers who understand humanistic content will automatically teach it, a notion that has sustained many inservice workshops since the 1960s.

In a related study, Kleine (1997) conducted a qualitative investigation into the inclusion of humanistic content in the taught curriculum of four science teachers with university undergraduate degrees in history, philosophy, botany, and education. In spite of the teachers' diverse academic backgrounds, Kleine could not detect, under close scrutiny, any differences among the teachers' humanistic taught curriculum, although she did note that all teachers' humanistic understandings were often relegated to a secondary consideration in their decisions about what to teach. Perhaps the four teachers were overwhelmed by primary considerations (listed earlier) or by the culture of their school.

In short, teachers' humanistic conceptual understanding does not necessarily influence classroom practice—the taught curriculum. To ameliorate this problem, Monk and Osborne (1997) proposed a pedagogical model that integrates humanistic content with teachers' main aims for teaching science, but the model has not been tested empirically at the present time.

Let me conclude by commenting on the research paradigms that guided the investigations cited above. The interpretive (qualitative) studies tended to provide rich in-depth data situated in a particular context that allowed the reader to consider complex relationships and subtle nuances and qualifications (e.g., McGinnis & Simmons, 1999; Mitchener & Anderson, 1989; Pedretti & Hodson, 1995), features necessarily overlooked in quantitative studies. As an illustration, Carlone (2003) presented qualitative evidence (discussed in detail later) to explain paradoxes found in her ethnographic study of teachers of a humanistic physics course, paradoxes such as the fact that stakeholders celebrated its legitimacy as an innovation but curtailed its growth within the school. Explanations for paradoxes do emerge from quantitative studies as well, but they do so as mere speculations to explain the variance in the data rather than as evidence-based findings reported in an interpretive research study.

The quantitative studies synthesized above often selected samples that were either very narrow or nonrandom, thus compromising their generalizability. Some quantitative studies, however, purposefully selected a narrow sample (e.g., from one particular educational jurisdiction), perhaps for the purpose of influencing policy makers or administrators within that educational jurisdiction—an instance of good politics of research.

SUCCESS AT IMPLEMENTATION

Successful implementation of humanistic science teaching has indeed occurred, though only under favorable circumstances. Success seemed to be associated with teaching grades 7 to 10 rather than higher grades, perhaps because teachers in these lower secondary grades were not confronted as much with the litany of obstacles to implementation listed earlier. Action research studies have been consistently successful at instigating teacher change toward a humanistic approach, perhaps because of the studies' relatively high proportion of human resources for the small number of participating teachers, and because of their relatively high proportion of eager volunteer participants (humanistic science teachers).

Research has identified numerous favorable circumstances, any combination of which has led to success. These are listed here: involvement of teachers in policy and curriculum development; involvement of teachers in producing classroom materials; establishment of supportive networks of teachers that included teachers experienced with humanistic science teaching who take leadership roles; a predisposition toward exploring new avenues of pedagogy and student assessment; a willingness to deal with degrees of uncertainty in the classroom; a substantial inservice program offered over a long period of time, coordinated with preservice methods courses and student teaching (internship) where possible; teacher reflection via diaries or journals and discussion; a recognition by teachers of the rewards from becoming socially responsible in their community, from enhancing their curriculum development and writing skills, and from improving their vision of science teaching; a responsive and caring project staff to provide the top-down guidance for achieving a balance with grassroots initiatives; contact with working scientists who convey intellectual, moral, and political support; an openness to evidence-based decisions founded on formative assessment and classroom experiences; and, finally, a focus on individual, autonomous, professional development into becoming, for example, a lifelong learner rather than a source of all knowledge in one's classroom (Anderson & Helms, 2001; Bartholomew et al., 2004; Briscoe, 1991; Cho, 2002; Eijkelhof & Kapteijn, 2000; Fensham & Corrigan, 1994; Gray, 1999; Gray & Bryce, 2003; Hart, 1989; Hart & Robottom, 1990; Keiny, 1993, 1996; Kilian-Schrum, 1996; Kortland, 2001; Ogborn, 2002; Osborne, Duschl, et al., 2003; Pedretti & Hodson, 1995; Reis & Galvão, 2004; Roberts, 1988, 1998; Rubba & Harkness, 1993; Sáez & Carretero, 2002; Solomon, 1999a; Tal et al., 2001; van Driel, Beijaard, & Verloop, 2001; Wang & Schmidt, 2001; Yager & Tamir, 1993).

Teachers in one study summarized their achievements realistically and metaphorically by describing "progress as a tension between a blessing and a curse" (Keiny, 1999, p. 347). The development of leadership capabilities among teachers makes a significant difference to the success at implementing a humanistic science program (Hofstein, Carmi, & Ben-Zvi, 2003), as does the acquisition of three types of knowledge: examples, anecdotes, and analogies related to the humanistic content (Ryder, Hind, & Leach, 2003b). Success at implementation must realistically be defined not in terms of 100% teacher adoption but in some practical proportion of teachers who shift their orientation significantly toward a humanistic perspective. For example, the act of implementing a humanistic science textbook in the United States stimulated and sustained changes to the teaching practices of one-third of a cross-section of teachers in a study by Ratcliffe (2004). This result may represent substantial success, relatively speaking. More research is needed.

Changing Teachers' Orientations

Large-scale research projects hold promise for supporting teachers' attempts at transforming their science teaching into a humanistic perspective. These have included the Iowa Chautauqua Program (Yager, 1996; Yager & Tamir, 1993); the Science Education for Public Understanding Program (Ratcliffe,

2004; SEPUP, 2003; Thier & Daviss, 2001; Thier & Nagle, 1996); the United Kingdom Science for Public Understanding syllabus (Osborne, Duschl, et al., 2003); the Dutch Public Understanding of Science curriculum (De Vos & Reiding, 1999; Eijkelhof & Kapteijn, 2000); and the Israeli Science, Technology, Environment in Modern Society curriculum (Dori & Tal, 2000; Tal et al., 2001). Individually these projects exemplify the many, empirically derived, favorable circumstances (listed earlier) that influence a teacher's orientation to a humanistic perspective on school science. The projects sustained the work of committed teachers who face continuous pressure to return to the status quo (Pedretti, 2003).

One in-depth research study offered particular insights into features of some middle-of-the-road teachers. Bartholomew et al. (2004) in the United Kingdom monitored and supported 11 volunteer teachers who were interested in implementing the United Kingdom national science curriculum's "ideas about science," specific ideas empirically derived from a large Delphi study (Osborne, Collins, et al., 2003), reviewed in Chapter 3. The researchers were interested in "what it *means* to integrate teaching about the nature of science, its practices and its processes, with the body of canonical content knowledge in a way which reinforces and adds to the teaching of both" (p. 11, emphasis in original). The researchers identified five "dimensions of practice," each of which consists of two extreme positions (i.e., two teacher orientations) that characterize less successful and more successful teachers, respectively:

1. Teachers' knowledge and understanding of humanistic content: from "anxious about their understanding" to "confident that they have a sufficient understanding."
2. Teachers' conceptions of their own role: from "dispenser of knowledge" to "facilitator of learning."
3. Teachers' use of discourse: from "closed and authoritative" to "open and dialogic."
4. Teachers' conception of learning goals: from "limited to knowledge gains" to "includes the development of reasoning skills."
5. The nature of classroom activities: from "student activities are contrived and inauthentic" to "activities are authentic and owned by students."

These dimensions of practice are not mutually exclusive. Nevertheless, they do help to detail changes in teachers' orientations to a humanistic perspective, more so than vague feelings of comfort or discomfort reported in the literature.

Small-scale but in-depth studies have also provided our research community with fruitful methodologies and encouraging findings. Several studies are mentioned here. Working closely with 14 secondary teachers, Luft (2001) found their beliefs and practices related to student-centeredness had changed differentially during her study, depending on the experience of the teacher: Neophyte science teachers "changed their beliefs more than their practices, whereas the experienced teachers demonstrated more change in their practices than their beliefs" (p. 517). Geddis's (1991) case study of a teacher introducing controversial issues into his science class traced how the classroom discourse improved the more the teacher paid attention to the ideology of the knowledge presented, the ideology inherent in the teacher's instruction, and the intellectual context of that instruction. Briscoe (1991) conducted a case study of one teacher's experience of changing his beliefs, which resulted in his changing the metaphors that described his teaching role (i.e., his image or vision of teaching), and thus changing his classroom practices. Briscoe concluded that teachers "need time to reflect on their own practices, assign language to their actions, and construct new knowledge which is consistent with the role metaphors they use to make sense of changes in their practice" (p. 197). These small-scale studies indicate the complexity of science teachers' shifting their orientation toward a humanistic perspective.

On a larger scale, Kilian-Schrum (1996) investigated (through interviews, classroom visits, and a questionnaire) what 400 teachers were going through as they attempted to implement an STS curriculum in Alberta, Canada, supported by authorized textbooks produced specifically for the curriculum. She concluded that one's self-image as a science teacher and one's loyalty to a scientific discipline both had to change before a teacher's taught curriculum approximated the humanistic intended curriculum. To examine this similarity between the intended and taught curricula more closely, McClelland (1998) and Jeans (1998) videotaped lessons of a subsample of 12 of these Alberta teachers as they taught what they felt was a humanistic science lesson that integrated STS and canonical science content. The videotapes were analyzed and each teacher was then interviewed about the lesson and about the researchers' interpretations of the lesson. By focusing on classroom events in the videotapes, Jeans (1998) was able to extend Briscoe's (1991) research methods and identify each teacher's image (or vision) of humanistic science teaching *in action*, as opposed to an image arising from interviews alone. Both McClelland (1998) and Jeans (1998) were aided in their analysis by an eight-category scheme I devised (Aikenhead, 1994d) for describing STS content and its combination with canonical science content. This scheme, which indicates the degree of importance afforded humanistic content (from low to high importance), is summarized here and then used to summarize McClelland's and Jeans's research findings.

The low-importance end of the eight-point scheme has three categories that do not alter the traditional scientist-oriented structure of a curriculum: (1) a little STS content for motivational purposes only, (2) casual infusion of more STS content but with no coherent purpose, and (3) a purposeful infusion giving even more time to STS content. However, category 4 integrates the two types of content (canonical and STS content) in a *student-oriented* fashion, but only within a single scientific discipline. A category 5 STS curriculum integrates both types of content in a student-oriented fashion, but it also integrates scientific disciplines as required by the humanistic context. The proportion and importance of STS content increases with ensuing categories until category 8, which most science teachers would view as social studies with a little canonical science thrown in.

Jeans (1998) was able to locate his teachers along this eight-point continuum (e.g., teachers were at categories 2, 3, 5, and 6) and to compare their "images in action" with the intended curriculum's image designated as category 4 in my scheme (Aikenhead, 1994d). Jeans's research included videotaped micro-teaching by preservice science teachers (reviewed later in "Preservice Experiences"). The research studies by McClelland (1998) and Jeans (1998) empirically validate two broad conclusions: The phrase "successful implementation of a humanistic perspective" has many meanings for teachers, and these meanings can be described by various schemes.

Mitchener and Anderson's (1989) case study of 14 teachers implementing a humanistic science curriculum explained how decisions to accept, alter, or reject the new course were all made on the basis of the *same* set of concerns. Five were identified: concerns over reduced canonical science content, discomfort with small-group instruction, uncertainties over student assessment, frustrations with the nonacademic type of student attracted to the new course, and confusion over the teacher's role. The researchers could not distinguish between the accepting teachers and the rejecting teachers solely on the basis of the teachers' concerns, yet the teacher orientation of both groups obviously differed.

Two very different case studies of innovative teachers documented how teachers coped with negative reactions from their colleagues, administrators, and parents, and how this coping affected the extent to which the teachers' taught curriculum matched the intended humanistic curriculum (Carlone, 2003; Lantz & Kass, 1987). Some teachers were successful, some were not.

Student Influences

Carlone's (2003) study reinforced earlier research findings that showed that students significantly exerted positive or negative influences on their teacher's decision to implement a new approach (Anderson & Helms, 2001; Désautels, Fleury, & Garrison, 2002; Kilian-Schrum, 1996; McRobbie & Tobin, 1995; Nicol et al., 2004; Thomas, 1999; Tobin, Seiler, & Walls, 1999; Tsai, 2000; Wallace & Louden, 1998). Researchers strongly recommended that innovators take into consideration students' ideas, goals, and conventions when innovators expect student roles to change—for example, from passive to active or from playing Fatima's rules to engaging in critical thinking or meaningful learning. Students need to buy into a humanistic innovation for it to be successful. Many students will, but some will not (Chapter 6).

Probably the most encouraging support for teachers to change their orientation has been students' highly positive reaction to humanistic science classes. Osborne, Duschl, et al. (2003) pointed out a key finding in their research into a United Kingdom humanistic syllabus: "Teaching a course which is enjoyed by students is . . . much more engaging and motivating for the science teacher" (p. 19).

Methods of Instruction

One central theme to the successful implementation of a humanistic science course is a teacher's instructional methods. Compared with traditional science teaching, a humanistic approach demands a wider repertoire of strategies such as divergent thinking, small-group work for cooperative learning, student-centered class discussions of scientific or social issues (some controversial, some not), use of daily media resources, hands-on community investigations, simulations, role playing, and decision making. Collaborative small-group discussions are widely used in humanistic science classrooms and students often respond capably (Kolstø, 2001a; Nicol et al., 2004; Ratcliffe & Grace, 2003; Solomon, 1988a). In an analysis of this wide repertoire of teaching methods, Byrne and Johnstone (1988) concluded that any method that engages students in an interactive way will make school science more relevant to them. Thus it is the quality of interactivity that is important, not a particular teaching method. A splendid lecture, for instance, might achieve interactivity. In a case study of two exemplary, humanistic chemistry teachers, Garnett and Tobin (1989) discovered that in spite of the teachers' similar humanistic orientation, each used distinctly different teaching strategies—whole-group versus individualized instruction.

Recently researchers have focused attention on using drama to ensure interactivity in humanistic science courses (Marchetti & Camino, 2003; Ødegaard, 2003a, 2003b; Simonneaux, 2001; Watts et al., 1997). Although studies support its value to student learning, cultural sensitivity is needed in non-Western cultures if, for example, authoritarian values predominate (Jegede, 1995; Ninnes, 1995). Similarly, research in multicultural settings reinforces Byrne and Johnston's (1988) findings about interactivity; however, the classroom environment must reflect the culture of the community (Brown, 2004; Hines, 2003; Hughes, More, & Williams, 2004).

A key question related to instruction concerns the sequence in which ideas are taught: canonical science content before humanistic content, or vice versa? Years of research by the Dutch PLON project produced the following five-phase sequence for best practice (Eijkelhof & Kortland, 1987; Eijkelhof & Lijnse, 1988). Begin with a societal event or issue that establishes an interesting authentic context (e.g., a court case on drinking and driving—phase 1). This usually creates the need to know a certain technology (e.g., a breathalyzer—phase 2), which in turn motivates students to engage with canonical science concepts and processes broadly relevant to the societal event and related technology (e.g., solution mixtures—phase 3). This knowledge is then used to better understand the original technology or more sophisticated technologies (phase 4). Phase 5 returns students to the original event or issue for which a thoughtful decision is often made, informed by appropriate humanistic, scientific, and technological content. (Chapter 6 synthesizes related research on student learning.)

COMPONENTS TO A TEACHER'S ORIENTATION

To give clarity to the holistic complexity of life in a science classroom, researchers have attempted to articulate science teachers' beliefs and values and to recognize contextual features of teachers' actions in the classroom. One key feature is the social system of the school. On the basis of 11 case studies conducted across the United States, researchers Stake and Easley (1978) concluded, "What [science] teachers do with subject matter is determined by how it sustains and protects them in the social system [of the school]. Subject matter that did not fit these aims got rejected, neglected, or changed into 'something that worked'" (p. 16.21). Other key contextual features include a teacher's wealth of past experiences (growing up at home, studying at university, etc.), the science classroom (the students, resources, etc.), and a teacher's images of good science teaching (Aikenhead, 1984; Jeans, 1998; Kleine, 1997; Munby et al., 2000; Roberts, 1998; Russell & Munby, 1991).

Tobin and McRobbie (1996), for instance, focused their research on teachers' actions that were determined by a teacher's beliefs, behaviors, and the context of that action. Krull-Romanyshyn (1996) drew upon a large-scale STS inservice project in Alberta (170 teachers) to delineate a science teacher's "functional paradigm" and to discover the successful strategies that caused a shift in a teacher's functional paradigm. Duffee and Aikenhead (1992) extended these ideas by proposing a heuristic model for "teacher practical knowledge" (TPK) to explain in a holistic way the complexity of classroom life. They assumed that teacher behavior arises from decisions made consciously or unconsciously based on a teacher's practical knowledge. These decisions result from many interacting sets of personal ideas (i.e., practical principles, rules of practice, values, and beliefs) that integrate a teacher's past experiences (teaching experiences, educational experiences, etc.) and a teacher's worldview. These personal ideas interpret the current teaching situation (the intended curriculum, specific students in the classroom, etc.) before a decision for action is made. All these considerations are filtered through a teacher's vision or image of how teaching should be, before a final decision for action is taken. Most decisions reflect this image (Briscoe, 1991; Hand & Treagust, 1997; McGinnis & Simmons, 1999; Russell & Munby, 1991). In short, TPK places emphasis on the individual teacher as decision maker influenced by a myriad of practical and cultural considerations. Clandinin and Connelly's (1996) narrative inquiry is an appropriate research methodology.

In a multicultural context, Atwater and Crockett (2003) interpreted science teachers' actions in terms of their "educational worldview," which is created from family, community, and past schooling experiences, all seen through the lens of cultural self-identity (Chapter 7).

Barnett and Hodson (2001) expanded TPK into "the work culture of teachers, derived from their roles as institutional, social, and political actors" (p. 433). They coined the label "teacher *context* knowledge" (TCK) to draw attention to "what good teachers know, do, and feel is largely about teaching and is situated in the minutiae of everyday classroom life" (p. 436). TCK emphasizes the institutional, social, and political contexts that influence teacher behavior.

By identifying the complexities inherent in a teacher's orientation to humanistic science teaching, heuristics such as TPK, educational worldview, and TCK compel science education researchers to consider a rich analysis of teachers' orientations toward changing the traditional science curriculum into a humanistic one, a change for most teachers as painful and personally challenging as a Kuhnian paradigm shift is for most scientists. Therefore, if a middle-of-the-road teacher is to develop a humanistic orientation to science teaching, many personal deep-seated traits, values, beliefs, and conventions must change. To quote a teacher in Roberts's (1998) Science Teachers Thinking study, "We had to learn a whole new way of teaching." And this must take place supported by the teacher's school and community (Aikenhead, 2000b; Carlone, 2003; Rigano & Ritchie, 2003; Ryder et al., 2003a; van Driel et al., 2001). For pipeline-enthusiast teachers to change their TPK or TCK, they would have to change their personality or worldview. This will likely not happen.

In retrospect, it seems sensible that many research studies into the implementation of a humanistic science curriculum selected humanistic science teachers to establish that such a curriculum could be implemented (i.e., it had credibility). Only when research involved middle-of-the-road teachers and pipeline enthusiasts did major problems arise that forced researchers to produce richer data, analyzed in greater depth. For instance, in their study into fostering the use of media reports related to socio-scientific issues, Ratcliffe and Grace (2003) concluded, "Only if the materials have some coincidence with teachers' interests and goals are there real attempts at incorporation into lessons, despite the support and encouragement offered" (p. 107).

One simple solution to the challenge of implementing a humanistic science curriculum may seem obvious: Preservice teacher education programs need to produce teachers who embrace humanistic perspectives. Researchers have explored that solution.

PRESERVICE EXPERIENCES

As with inservice studies, research into preservice science teachers' orientation to a humanistic perspective first focused on documenting students' understanding. Many students expressed naïve and simplistic ideas (Cunningham, 1998; Nieswandt & Bellomo, 2003; Tobin, Seiler, & Smith, 1999; Tsai, 2001a), but some expressed contemporary and complex ideas (students with strong academic backgrounds in the history, philosophy, and sociology of science, and those with scientific experience in industry and government labs). The research evidence showed that this understanding *did* make a noticeable difference in practice-teaching (internship) settings generally supportive of such innovations (Bianchini et al., 2003; Cunningham, 1998; David, 2003; Nieswandt & Bellomo, 2003; Schwartz & Lederman, 2002).

However, several studies reported contradictory findings. Jeans's (1998) analysis of 35 videotaped STS microteaching lessons clearly indicated that his preservice teachers were not appreciably including their humanistic ideas in their lessons. He concluded that these preservice teachers mimicked the canonical science content orientation of their recent university science classes and succumbed to peer pressure to demonstrate subject matter expertise. David (2003) and Schwartz and Lederman (2002) discovered a different reason to explain the reluctance of preservice teachers to include humanistic content in their student-teaching lessons: Novice teachers naturally lack confidence in teaching canonical science content, and until a reasonable confidence is attained, humanistic instructional outcomes are relegated to a low priority. Background knowledge of humanistic content seems to exert an influence in some preservice settings but not in others.

The transition from a preservice humanistic science education program to first and second years of teaching showed similar results. "In spite of preservice students' rather sophisticated views about the nature and practice of science, and strong endorsements of STSE [science-technology-society-environment] education, very little materialized by way of practice during their first year of teaching" (Pedretti, 2003, p. 230). One way to strengthen the transition is to engage preservice students in critically reflecting on personal factors that tend to inhibit a humanistic perspective. This was accomplished in a methods course by an interactive multimedia case of a successful humanistic science teacher (Pedretti, Bencze, Hewitt, Jawani, & Romkey, 2004). As a result, students identified and discussed their own conflicts of practice and ideology. They agonized over teacher authority (i.e., professional identity), over losing the support of their colleagues, over covering the curriculum (i.e., discipline loyalty), over political issues in the classroom (i.e., images of science), and over expressing biases—all concerns that were identified earlier in this chapter. A follow-up longitudinal study is under way.

Researchers have assessed university courses that purport to transmit a contemporary understanding of humanistic content to preservice science teachers. Their disappointing results show that university students generally do not easily reconstruct their naïve and simplistic preconceptions into contemporary conceptions (Abd-El-Khalick & Lederman, 2000; Cunningham, 1998; Gallagher, 1991; Lederman, 1992), likely because those preconceptions are anchored in personal beliefs, values, ideologies, identities, allegiances, and goals. The university model of instruction, the transmission model, is not very effective for this purpose (Bencze & Hodson, 1999; Rubba & Harkness, 1993; Solomon, 1999a). Changing personal deep-seated ideas about humanistic and science content often requires much more than a methods course; it takes a whole university (Abd-El-Khalick & BouJaoude, 1997; Cunningham, 1998; Nieswandt & Bellomo, 2003). For cross-cultural science teaching, success also seems to require a self-reflection by preservice teachers on their self-identities (Barker, 2004; Chinn, 2004).

Research into successful approaches to teaching humanistic science content to preservice teachers strongly suggests that these experiences be experiential, reflective, collaborative, and critical (Bencze & Hodson, 1999; Bianchini & Solomon, 2003; Schwartz, Lederman, & Crawford, 2004; Solomon, 1999a). The experiences should address various aspects of teacher development (professional, social, and personal) and should be guided by heuristics such as teacher practical knowledge (Duffee & Aikenhead, 1992), educational worldview (Atwater & Crockett, 2003), or teacher context knowledge (Barnett & Hodson, 2001). Lin's (1998) research into training preservice teachers on how to teach chemistry through the history of science provides both quantitative and qualitative data. Modest gains on quantitative measures were interpreted through interview data to support the conclusion that when education students understand the history of science, they tend to use that knowledge rather than their preconceptions when they discuss humanistic content in their science classes.

The most influential experience for preservice teachers is student teaching (internship) where apprentices work with an experienced teacher (Roth, Tobin, Carambo, & Dalland, 2004; Russell, McPherson, & Martin, 2001). Only when novice teachers are placed in supportive apprenticeships can their humanistic perspective develop further (Bianchini & Solomon, 2003; David, 2003; Nieswandt & Bellomo, 2003; Tsai, 2001a); otherwise pipeline enthusiasts and middle-of-the-road teachers direct their apprentices to forget the ivory tower humanistic approaches to school science presented in their preservice courses (Abd-El-Khalick et al., 1998; Barnett & Hodson, 2001; Munby et al., 2000; Russell et al., 2001). And thus the vicious cycle reproduces the status quo—another political reality that challenges the educational soundness of a teacher education program.

SCHOOL CULTURE

In research related to the taught curriculum, tension was evident between educationally sound arguments for a humanistic perspective, on the one hand, and the political reality of institutional expectations, customs, ceremonies, beliefs, and loyalties (i.e., school culture) favoring the status quo, on the other.

Inservice and preservice humanistic science projects have largely failed to achieve the radical changes in school science envisioned by their project leaders. When analyzing their failures, many researchers chose not to consider power and political reality. Instead they focused on crafting educationally sound statements in the form of nonpolitical ideas and recommendations for teacher development. Some researchers, however, have

placed political reality on their agendas because the enactment of an intended humanistic science curriculum not only takes place with individual
teachers and their unique orientations to humanistic school science; it also
takes place within a political arena of students, colleagues, administrators,
the school, and the immediate and extended community (Aikenhead,
2000b; Bianchini & Solomon, 2003; Carlone, 2003; Fensham, 1992; Gaskell,
1989, 2003; Medvitz, 1996; Munby et al., 2000; Pedretti & Hodson, 1995;
Roberts, 1988). This political arena has been researched explicitly and has
yielded a treasure of evidence concerning the culture of school science and
the culture of typical schools.

Humanistic science is integrative by nature. An everyday event usually requires the integration of several disciplines to understand it adequately, and, in turn, these disciplines themselves are integrated with
human features of the event (e.g., with technology, social systems, and
culture; Hansen & Olson, 1996). Venville et al.'s (2002) review of curriculum integration sheds light on school culture in which *status* is a principal
factor. Status in most school cultures is *high* for courses that are rigid in
their course content, highly differentiated and insulated from other subjects, and academically and idealistically objective. On the other hand,
status is *low* for courses that are flexible in their content to achieve relevance and timeliness; amenable to overlap with other subjects; and utilitarian, relevant, and subjective. Status is animated by the language used
within a school, for example, "hard" and "soft" sciences, indicating high
and low status, respectively. Clearly, humanistic science courses currently
fall in the low-status category, and this directly affects who teaches them
(Carlone, 2003; Gaskell, 1989), which in turn sustains their low status.
For instance, teachers with a general science background who normally
taught home economics and technology courses were recruited to teach
the humanistic "Science & Technology 11" course in British Columbia,
Canada, largely because many regular high school science teachers refused
to take it on (Gaskell, 1989). Consequently, a low status was quickly conferred on this innovative curriculum in many schools, in spite of the endorsement it received from the provincial Ministry of Education, from
David Suzuki (a renowned television scientist), and from the Science
Council of Canada, all high-status agencies.

Roberts's (1988) research into the politics of the science curriculum recognized the roles of status and loyalty within the culture of school science.

If one wants to promote science teacher loyalty to a science curriculum proposal: guarantee the status of the content by enshrining it in an acceptable,
recognized examination, and secure the support of the subject community.
Otherwise the specter is ever present, for the teachers, that the proposal's

academic status will degenerate to utilitarian and pedagogic limbo. . . . [Loy-
alty] is quite a different matter from the need for in-service education to
ensure that the teachers *understand* a new proposal. (p. 48, emphasis in
original)

Teacher loyalty forms a bridge between teacher orientation and school
culture.

In the United States, Carlone (2003) conducted a highly informative,
ethnographic case study of this bridge, showing specific tensions between
support for, and constraints on, a humanistic physics course that used the
textbook *Active Physics* (Eisenkraft, 1998). The study took place in a large,
upper-middle-class high school proud of its graduates' postsecondary enrollment figures, proud of its five different grade 11 and 12 physics courses
each with multiple sections, and proud that half its students enrolled in
the school's first-year physics program. *Active Physics* was one of three grade
11 physics courses offered at this school. Backed by the status of both the
American Association of Physics Teachers and the American Institute of
Physics, *Active Physics* contextualizes physics content in social issues, in
technology-centered lab activities (i.e., tool use), and in everyday events.
In other words, the course is a combination of functional, personal curiosity, need-to-know, and wish-they-knew science as described in Chapter 3. Instruction emphasized a social constructivist view on learning. The
course's two well-credentialed, humanistic physics teachers (Ms. Carpenter
and Mr. Stewart, a minority among 11 physics teachers) had initiated the
course and had taught it 4 years, successfully preparing students for grade
12 advanced placement physics. By Roberts's (1988) standards (quoted
earlier), the course met the conditions for being enshrined as acceptable
in this school.

Not only does Carlone's (2003) case study illustrate teachers' decisions
to implement or not to implement a humanistic perspective (synthesized
earlier), but also it clarifies how status is dynamically and concretely played
out and negotiated in this school's culture. For example, "Ms. Carpenter
said that she thought she had a large role in maintaining the survival of
Active Physics because she was more 'political' than Mr. Stewart. She did
more of the public relations work with *Active Physics* in that she spent more
time trying to convince others of its legitimacy" (p. 322). "Interestingly,
this demonstration of legitimacy was enough to ensure the survival of *Active
Physics*, but not enough to ensure its growth and Mr. Stewart and Ms. Carpenter's prestige within the department" (p. 326). Many reasons accounted
for this, but only a few are summarized here. Neither teacher had political access to the many students in regular physics classes who seemed to
garner personal status as academic students by pejoratively calling *Active*

Physics "blow up" (easy) physics. The teachers of regular physics protected their own superior position in the school hierarchy of status in a variety of ways: by rationalizing (e.g., we must protect the sanctity of physics), by belittling (e.g., calling *Active Physics* "grade 9 science"), and by marginalizing (e.g., the professional interaction between *Active Physics* teachers and the other teachers was restricted). Although the administration proudly provided substantial financing to implement this tool-centered humanistic course, it restricted the number of sections offered and thus it did not allow the course to expand. Nor did the administration provide sufficient political support to ameliorate the isolation between the innovators and the other physics teachers. Perhaps the administrators were simply balancing the politics of the school in two ways: first, wanting to look innovative to the public by offering a radically different physics course but, at the same time, maintaining an aura of academic (traditional) excellence by ensuring most students took regular physics; and second, balancing teacher loyalties to the radically new course and to the status quo courses. Carlone (2003) pointed out, however, that this isolation provided Mr. Stewart and Ms. Carpenter "the freedom to enact their visions of good science education without having to coordinate their curriculum's content and methods with other teachers who may have had different ideologies" (pp. 325–326).

In Rigano and Ritchie's (2003) case study of the head of a junior secondary science department, the researchers described how his attempts to implement a progressive state curriculum challenged the school culture's emphasis on public image, parental demands, and traditional ideologies. Three main barriers to his attempts were the backgrounds of fellow staff (mostly pipeline enthusiasts or middle-of-the-road teachers entrenched in comfortable ways), school expectations (transmission instructional methods, traditional student assessment, and a reliance on prescribed textbooks), and the individual isolation of teachers' classroom practice (insulated from each other by "norms of egalitarianism, privacy, politeness and contrived collegiality" [p. 313]). This school culture could only be renegotiated through developing shared beliefs throughout the department and the school. The department head's limited incremental success was attributed to his constant championing the voice of students. Fundamental to teachers' orientations to school science is the desire to do what is right for students.

These case studies detail the political challenges facing educators wishing to implement a humanistic approach in school science. The case studies join with other research findings (Aikenhead, 2000b; Gaskell, 2003; Medvitz, 1996; Osborne, Duschl, et al., 2003; Roberts, 1988) in concluding that the powerful 19th-century legacy of school science can be challenged successfully on a small scale. However, challengers must be prepared to renegotiate the culture of school science and to renegotiate some social structures of privilege and power along the way. The innovators in the case studies became politicized at the school level, but that was not sufficient to ensure that the innovation, however educationally sound, would affect the whole science program. Moreover, they simply did not have the time and energy to become politicized at the regional or nation level, as Pedretti and Hodson (1995) proposed teachers do.

CONCLUSION

Elmore (2003) drew upon a great deal of research and experience with school innovation when he cryptically characterized a typical science education innovation study as follows: a gathering of "the faithful" (i.e., humanist science teachers) to show that the innovation can work on a small scale, and then leave "the virus" (i.e., the innovation—a humanistic approach) to populate the system on its own because the innovation is such a good idea (i.e., educationally sound). This approach to changing school science has continually failed, mostly as a result of problems arising when the researchers attempt to transfer the innovation from the research participants (humanistic teachers) to other teachers (pipeline enthusiasts and middle-of-the-road teachers), a move Elmore called "scaling-up." The attempt to scale-up the innovation so more teachers will use it usually creates ideological conflicts between humanistic teachers and pipeline enthusiasts. I must agree with Elmore. The majority of the research literature on teacher orientation to a humanistic science curriculum is comprised of small-scale studies with a few volunteer science teachers to initiate the novel project, projects that did not have sufficient resources to expand in scale or to be sustained over time (Anderson & Helms, 2001).

   To meet the challenge of implementing humanistic school science on a large scale, we require an actor-network (Gaskell & Hepburn, 1998) larger than two teachers, peer coaching or professional development internships, collaborative action research, and case studies of successful implementation for teachers to read. Political reality dictates that an expanded actor-network needs to be formed in concert with socially powerful groups, for example, a school system's administration (Elmore, 2003), or a much more pervasive group such as local or national industries and corporations (Dori & Tal, 2000; Gaskell, 2003). The challenge of enacting a humanistic science curriculum at the school level comes down to issues of scale (e.g., moderate or ambitious); resources to engage in the appropriate politics of change (e.g., altering the curriculum's status perceived by stakeholders, altering teachers' loyalties, and altering the assessment system); finances and infrastructure (e.g., to support teachers and students, and to sustain their efforts over a long period of time); and, finally, availability of science teachers who already have, or will develop, an orientation to such a curriculum. These local issues must be placed in the broader context of curriculum policy development (Chapter 3) and placed on the political agenda to renegotiate the concept of status in school culture.

Teacher orientation to humanistic school science is a dynamic entity that interacts with, and on occasion is modified by, many influences that determine teacher development. One key influence is the impact the taught curriculum has on students' learning, a topic to which we now turn.

6

Student Learning

The learned curriculum, planned or unplanned, is given high priority in arguments concerning educational soundness. This chapter presents evidence from research into student learning in humanistic school science.

As evident throughout the book, humanistic science education has various learning outcomes. On the basis of the research literature, I summarize these outcomes into five interrelated categories: (1) to make the human aspects of Western science more accessible and relevant to students (e.g., its sociology, philosophy, and history, its interrelationships with technology and society, as well as its status as a subculture found worldwide); (2) to help students become better critical thinkers, creative problem solvers, and especially better decision makers in their everyday lives related to Western science and technology; (3) to increase students' capability to communicate with scientific and technological communities or media spokespersons (i.e., to listen, read, respond, etc.) in order, for example, to feel more at home in their own culture increasingly shaped by Western science and technology; (4) to augment students' commitment to social responsibility (citizenship); and (5) to increase interest and achievement in learning canonical science found in the traditional curriculum or in other sources of Western scientific knowledge, depending on the context.

Purposefully missing from this list of school-based outcomes are lofty and nonassessable aims such as an empowered citizenry, an enlightened democracy, and wise and responsible decision makers. Interestingly, however, science museums and science centers seek to assess the impact their exhibits have *on students' lives* (Dierking, Ellenbogen, & Falk, 2004), an outcome more ambitious than most schools dare assess. This goal exemplifies teaching science-as-culture content (Chapter 3). Research at science centers has documented the transformative impact on students of issues-based exhibits that engaged students in humanistic science content (Pedretti, 2004).

Researchers have operationalized and prioritized the five outcomes (listed above) in various ways. Their research into the learned humanistic science curriculum is synthesized here in the following sequence: canonical science content acquired by students in humanistic science courses, evidence-gathering techniques for assessing humanistic content, summative assessment in quasi-experimental studies, other investigations in humanistic science education, and student decision making.

CANONICAL SCIENCE CONTENT

One expectation of humanistic school science was to increase interest and achievement in canonical science content (expectation 5, above). Researchers anticipated significantly higher achievement in canonical science by students in a humanistic science course (Abd-El-Khalick & BouJaoude, 2004; Aikenhead, 1980; Eijkelhof & Kortland, 1987), yet these higher expectations rarely materialized (Eijkelhof & Lijnse, 1988; Irwin, 2000; Kortland, 2001; Welch, 1973). As discussed in Chapter 3, canonical science content is not often usable in everyday situations, and most students encounter extreme difficulty when they attempt to learn canonical science meaningfully at school, no matter how relevant the context (Eijkelhof, 1990; Hennessy, 1993; Lijnse, 1990; Osborne, Duschl, et al., 2003; Ramsden, 1997; Solomon, 1983). This finding was explained by Solomon (1988b) when she clarified how a relevant social context (in which the canonical content is embedded) provokes affective and value-laden connections in students' minds, thereby making the situation far more complex to think through, especially for less able students. Research into students' socioscientific decision making (reviewed later in this chapter) corroborates and articulates Solomon's claim further.

A paradox presents itself: the greater the social or cultural relevance associated with canonical content, the greater the student *motivation*, but the greater the *complexity* to learn it meaningfully.

These two factors (motivation and complexity) may cancel each other out, yielding the *equal achievement* in learning canonical science content, as was found in many research studies that compared students enrolled in humanistic science courses with their counterparts in traditional ones (Aikenhead, 1994b; Bybee, 1993; Cho, 2002; Eijkelhof & Lijnse, 1988; Galili & Hazan, 2001; Irwin, 2000; Klopfer & Cooley, 1963; Ramsden, 1997; Seker, 2004; Welch, 1973; Winther & Volk, 1994). However, motivation can overcome complexity and lead to greater achievement on traditional science tests favoring students enrolled in humanistic science courses (Häussler & Hoffmann, 2000; Lin, Hung, & Hung, 2002; Lubben & Campbell, 1996; Mbajiorgu & Ali, 2003; Rubba & Wiesenmayer, 1991; Sadler, 2004; Solomon et al., 1992; Solomon, Scot, & Duveen, 1996; Sutman & Bruce, 1992; Wang & Schmidt, 2001; Wiesenmayer & Rubba, 1999; Winther &

Volk, 1994; Yager, 1996; Yager & Tamir, 1993; Zohar & Nemet, 2001). Other research (Chapter 3) suggested that only when personal action is at stake (i.e., very high motivation) do most citizens work through the complexity to learn enough content to take action; but then the science content learned meaningfully is not likely to be the "pure science" found in the traditional curriculum (Jenkins, 1992; Lawrence & Eisenhart, 2002; Layton, 1991).

Scientists realize that their own fundamental presuppositions about the world must change dramatically before they can work through the complexity of a new paradigm replete with counterintuitive ideas. For example, physicist Greene (1999) concluded, in the context of explaining Einstein's special theory of relativity, "Understanding and accepting [counterintuitive ideas] requires that we subject our worldview to a thorough makeover" (p. 27). Because most students hold worldviews dissimilar to a scientific type of worldview (Cobern, 1996b, 2000), meaningful learning of science content requires students to undertake a thorough makeover of their worldviews. How many students will actually do this when they can play Fatima's rules instead?

Humanistic science educators generally take on an extremely high and inappropriate standard of excellence in learning canonical science content if they expect a higher degree of meaningful learning to occur when they embed that content in relevant everyday contexts.

On the other hand, when canonical science content was contextualized in local non-Western cultures, research documented increased interest and achievement by non-Western students (Barnhardt, Kawagley, & Hill, 2000; Jegede & Okebukola, 1991; O. Lee, 2002; Lubben & Campbell, 1996; Riggs, 2005). Many of these students experienced a culture clash between their everyday culture and the culture of school science (Chapter 7), a clash detrimental to their science achievement (Erickson, 2004; Jegede, 1995; Lee, 1997; Lynch, 1996; Maddock, 1981; Norman et al., 2001) and detrimental to valid assessment (Solano-Flores & Nelson-Barber, 2001), although not for every non-Western student (Koul, 2003). Motivated non-Western students seem to deal more effectively with culture clashes when Western science content is contextualized locally—another case of motivation overcoming complexity, perhaps.

EVIDENCE-GATHERING TECHNIQUES FOR HUMANISTIC CONTENT

Another expected outcome of humanistic school science is to make accessible the human aspects of science. Over the past 50 years, the evidence-gathering techniques for assessing the human aspects of science have developed dramatically, beginning with the quantitative paradigm (Aikenhead, 1973; Cheek, 1992; Lederman, 1992), moving to an interpretive (qualitative) paradigm (Aikenhead, 1974, 1988; Aikenhead, Fleming, & Ryan, 1987; Aikenhead & Ryan, 1992; Driver et al., 1996; Lederman, Abd-El-Khalick, Bell, & Schwartz, 2002; Wade, Lederman, & Bell, 1997), and on to a situated-cognition approach within the interpretive paradigm (Gaskell, 1994; Solomon, 1992; Welzel & Roth, 1998). The content of each instrument or protocol varies greatly from study to study. Content associated with the philosophy of science (i.e., epistemology, ontology, and some axiology) became known as the "nature of science" (Lederman, 1992), whereas content associated with various sociologies of science and societal contexts of science has had many labels, such as "the social aspects of science." Some researchers have expanded the category "nature of science" to include these social aspects of science (Lederman et al., 2002; McComas, 1998; Millar & Osborne, 1998). In this book, I treat all of these ideas as humanistic content.

More important, consensus has not been reached on what ideas represent the most acceptable or defensible views (e.g., Alters, 1997; Smith, Lederman, Bell, McComas, & Clough, 1997). Contested areas of scholarship remain (for instance, the realism-constructivism debate). The central issue here for researchers is the validity or trustworthiness of student assessment.

Many research studies have narrowly focused on only three or four humanistic ideas (owing to time and resource limitations), and these ideas have varied from study to study. Consequently, a plethora of instruments and protocols have been published over the years, representing the full range of contested and uncontested humanistic content. These instruments are not reviewed here (see, e.g., Abd-El-Khalick & Lederman, 2000; Aikenhead, 1973; Wade et al., 1997), but their diversity is noted to draw attention to the potential problem of inconsistency among research studies using different evidence-gathering techniques.

Researchers working within the quantitative paradigm talk about measuring students' attainment of humanistic ideas and about sorting students into categories of achievement (e.g., literate and illiterate). This paradigm is characterized by judgments based on external criteria, such as views of a panel of experts or a theoretical position accepted by scholarly academic communities. These quantitative instruments generally suffer from three types of problems. First, an ambiguity problem arises from researchers' erroneously assuming that their meaning ascribed to a statement is exactly the meaning read into the statement by students (Aikenhead, 1988; Aikenhead et al., 1987). Second, instruments suffer from a validity problem caused by the following:

- The "correct" responses are contingent upon the views held by the instrument's developer (Lederman et al., 2001); or
- A panel of experts decides what is correct even when the panel selection is biased or a low interjudge reliability is reported (Manassero-Mas, Vázquez-Alonso, & Acevedo-Díaz, 2001; Rubba, Schoneweg-Bradford, & Harkness, 1996; Vázquez-Alonso & Manassero-Mas, 1999); or
- The validity and reliability of instruments are not adequately established.

And third, there is a problem with the narrow scope of outcomes assessed by pencil-and-paper or multiple-choice instruments, outcomes that fail to capture the rich and diverse array of anticipated outcomes for humanistic science courses (Cheek, 1992).

Researchers working within the interpretive paradigm, on the other hand, are primarily interested in clarifying and understanding students' views and conveying those views to others. To accomplish this task, a variety of protocols have been developed (Aikenhead et al., 1987; Aikenhead & Ryan, 1992; Driver et al., 1996; Leach, Driver, Millar, & Scott, 1997; Lederman et al., 2002; Moss, Abrams, & Robb, 2001; Solomon, 1992; Solomon, Duveen, & Scot, 1994). The varied characteristics of these protocols are captured in the following three dichotomies:

1. Interviews (semistructured to nonstructured) versus written responses (open-ended to preestablished responses).
2. General questions posed without a context provided (e.g., Do scientists' personal values ever affect the research results they obtain?) versus contextualized questions embedded in clearly described situations or in a particular task undertaken by students (e.g., learning about a particular scientific controversy such as the dispute surrounding Wegener's continental drift hypothesis, and then discussing specific focus questions in small groups).
3. Questions that make the humanistic content explicit for students such that students knowingly discuss the explicit content (e.g., "Yes, scientific truth is what scientists decide is true by consensus") versus questions that leave the humanistic content implicit for students and this content is inferred by researchers who analyze student responses responses such as, "Scientists must publish their theory before it becomes accepted."

Two further characteristics of interpretive protocols in general should be added to this list and are also stated here as dichotomies for the sake of clarity:

4. Students' personal knowledge about school science versus their knowledge about science and scientists in authentic workplaces (Gaskell, 1992; Hogan, 2000; Larochelle & Désautels, 1991; Leach et al., 1997; Solomon et al., 1994).

5. Knowledge understood by students well enough to articulate it in an interview versus knowledge understood and believed strongly enough by students to be guided by it as they participate in a discussion or simulation (Dahncke, 1996; Driver et al., 1996; Gaskell, 1994; Solomon, 1992, 1994d; Welzel, & Roth, 1998).

Research studies have often used various combinations of these protocol characteristics to augment the trustworthiness of their data, and several studies have combined the quantitative and interpretive paradigms.

The fifth characteristic of interpretive protocols (stated directly above) identifies a situated-cognition approach within the interpretive paradigm. On the basis of their empirical evidence, Welzel and Roth (1998) questioned the assumption that interviews capture students' conceptual knowledge accurately. Three points were argued: First, interviews themselves are contrived because they are not situated in the context of action. Second, Welzel and Roth's data suggested that students' humanistic concepts are not highly stable from context to context, contrary to what science educators had assumed. Third, ambiguity can plague interviews, too. Thus, to understand what students have learned in humanistic science classes, researchers need to listen to student conversations, note the actions of students as they engage in a meaningful task, and interview them about that specific task. (This research technique is similar to analyzing videotapes of teachers *in action* rather than interviewing teachers *about* their recollection of actions, as mentioned in Chapter 5.)

Sutherland and Dennick (2002) and Sutherland (2005), for instance, illustrated Welzel and Roth's (1998) recommendation when they had First Nations students discuss concrete critical incidents about science, from which the researchers interpreted humanistic ideas about science held by those students. The researchers did this over a number of different contexts represented by different critical incidents. Their analysis found consistency among contexts. Critical incidents have been successfully used in other studies as well (Bell & Lederman, 2003; Leach et al., 1997).

Ethnographic researchers typically observe students during a school science activity and interpret their words and actions as manifesting certain humanistic concepts about science. Leach et al. (1997) criticized this research technique as having a limited school science focus, and cautioned that observers may be detecting sophisticated versions of Fatima's rules rather than detecting humanistic concepts guiding the students' personal practice.

Gaskell (1994) addressed these problems in a study in which students viewed commercial television clips of science-based controversial topics (e.g., sun-tan parlors) and engaged in free discussion in small groups. These discussions were subsequently analyzed. In a second phase of the study, students were presented with a related story that personalized the issue for them. They were interviewed on two occasions, once after the first presentation (television clips) and again after the personal story. During the second interview, Gaskell challenged students' original key ideas to determine how strongly students held them and what other ideas supported their original key idea. He reasoned, "It is in the dynamics of coping with challenges to their points of view that students articulate the array of elements that they associated with an issue and also the strengths and weaknesses of the links between the various elements or points" (p. 312). Gaskell's protocol yielded trustworthy data and high transferability to everyday events outside of school, and it afforded insight into citizens making a decision on a socioscientific issue (a principal component to a humanistic science curriculum, discussed below). On the other hand, the protocol is labor intensive and limited to a few issues found in the mass media.

There is no one best technique or instrument for gathering evidence; each has advantages and limitations. In the interpretive paradigm of research, for example, one of the most comprehensive protocols, Views on Science-Technology-Society, VOSTS (Aikenhead, Ryan, & Fleming, 1989), catalogued humanistic content found in contemporary literature and transposed it into 114 novel multiple-choice items developed collaboratively with students (Aikenhead & Ryan, 1992). Each item contextualized an issue about which students were asked to express a view plus identify their reason for holding that view. The development process itself established the instrument's trustworthiness, whereas its test-retest reliability was independently demonstrated (Botton & Brown, 1998). Although VOSTS was developed within the interpretive paradigm of research, marking schemes have been constructed for studies in the quantitative paradigm (Manassero-Mas et al., 2001; Rubba et al., 1996; Vázquez-Alonso & Manassero-Mas, 1999). But VOSTS is limited in at least three specific ways: Students typically can respond to only about 15 to 20 items in one sitting; VOSTS does not provide the flexibility to probe students' responses as interviews do; and, similar to a humanistic perspective itself, VOSTS is not universal and, therefore, when it is used in settings culturally different from Anglo Canada, researchers must empirically modify and validate the items for use in that culture (e.g., in Spain—Manassero-Mas & Vázquez-Alonso, 1998; in Taiwan—Lin, 1998; and in Nigeria—Mbajiorgu & Ali, 2003); otherwise, problems arise (e.g., in Lebanon—Abd-El-Khalick & BouJaoude, 1997).

Lederman et al. (2002) designed the Views of Nature of Science (VNOS), purposefully avoiding the response constrictions of VOSTS. VNOS-form C is a 10-item, open-ended questionnaire, of which most items are decontextualized. Part of the protocol for using VNOS-form C is a follow-up interview schedule to investigate students' views further, for instance, to discover how students spontaneously contextualized the items as students responded to them. Similarly, Driver et al. (1996) avoided the constrictions of VOSTS and developed a complex protocol (Images of Science Probes) that engaged students in a double task: A class presentation introduced students to a science-related dispute (e.g., the safety of irradiated food) and was followed by a small-group discussion of some key questions about the dispute. Researchers then conducted semistructured contextualized interviews with pairs of students to focus on general humanistic concepts related to science as a social enterprise, the nature or status of scientific knowledge, and the purposes of scientific work. Six probes allowed researchers to check for consistency of students' key concepts from one context to another. The research team developed interpretive grids to help future researchers interpret students' interview responses. Both the VNOS-form C and the Images of Science Probes provide flexible techniques for gathering evidence in depth, even though the approach is labor intensive and the breadth of topics is necessarily restricted to a few key humanistic concepts.

Evidence-gathering techniques have also included assessment embedded in the instruction using everyday events (Thier & Nagle, 1994; Wilson, 2004); formal case study assessment of student attitudes (Dori & Tal, 2000); formal compilation of relevant student actions within the parameters of computer simulations (Dahncke, 1996) or outside of school in the real world (Dori & Tal, 2000; Jiménez-Aleizandre & Pereiro-Muñoz, 2002; Rubba & Wiesenmayer, 1991; Wiesenmayer & Rubba, 1999); simple straightforward questionnaires developed by the researcher (Solbes & Vilches, 1997); and informal assessment via interviews with students and parents (Dori & Tal, 2000).

SUMMATIVE ASSESSMENT IN
QUASI-EXPERIMENTAL STUDIES

Researchers invariably design summative assessment studies by crafting educationally defensible research questions while at the same time considering the political context of their work (Welch, 1979). Given that humanistic perspectives are generally ignored or marginalized by traditional school science, researchers who embrace a humanistic ideology will often design their summative assessments to demonstrate the advantages of their innovation over the status quo, and thereby attempt to sway policy makers and science teachers oriented to the pipeline ideology. Because the target audience of this research tends to value experimental methods, a strong argument in favor of humanistic science education must be based on quantitative evidence derived from quasi-experimental research designs. This was particularly true of studies prior to 1990.

Since then emphasis in science education research has clearly shifted to interpretive studies (synthesized below). Perhaps our community heeded Welch's (1969) warning about the pitfalls of overly simplistic "horse race" evaluation studies (i.e., experimental versus control groups). It is interesting, therefore, that some major education initiatives out of Washington, D.C., have recently rejected all but quantitative ("scientifically based") research for political accountability (Eisenhart & Towne, 2003). Interpretive researchers have been more concerned with educational soundness achieved through understanding what is going on in humanistic science classes and in the minds and hearts of students.

By the 1900s it was evident in a literature review I conducted (Aikenhead, 1994b) and in two reviews of the extensive Iowa Chautauqua Program by Yager (1996) and Yager and Tamir (1993) that sufficient summative evaluation studies had been published to warrant the following research synthesis. Students in humanistic science classes (compared with those in traditional science classes) can

- significantly improve their understanding of social issues both external and internal to science, the nature of science, and the interactions among science, technology, and society; but this achievement depends on what humanistic content is emphasized and evaluated by the teacher. The teacher and overt instruction make the difference.
- significantly improve their attitudes toward science, toward science classes, and toward learning as a result of learning humanistic content.
- make modest but significant gains in thinking skills such as applying canonical science content to everyday events, critical and creative thinking, and decision making, as long as these skills are explicitly practiced and evaluated in the classroom.
- benefit from studying science from a humanistic perspective provided that the humanistic content is integrated with science content in a purposeful, educationally sound way; appropriate classroom materials are available; and a teacher's orientation toward school science is in reasonable synchrony with a humanistic perspective

(i.e., humanistic teachers and similar minded middle-of-the-road teachers).

Since 1990, some studies have used a quantitative quasi-experimental design and their findings have reinforced the points made just above (Galili & Hazan, 2001; Khishfe & Abd-El-Khalick, 2002; Schwartz-Bloom & Halpin, 2003; Seker, 2004; Solbes & Vilches, 1997; Wiesenmayer & Rubba, 1999; Zohar & Nemet, 2001).

In Taiwan, for instance, Tsai (1999, 2000) investigated the relative effect of STS and traditional science classrooms on students' cognitive structures. The results from a questionnaire and from a "flow map" analysis of in-depth interviews delineated specific benefits that accrued from the STS classes, including more "constructivist views of science" and a better understanding of "the importance of social negotiations in the scientific community and cultural impact on science" (1999, p. 1201). Furthermore, Tsai's (2000) research suggested that students' epistemic beliefs may influence students' receptivity to humanistic school science, thereby helping to explain an earlier conclusion in this book (Chapter 5) that students could inhibit the implementation of a humanistic approach to science teaching, especially students with a positivistic view of science.

According to a number of science education researchers (e.g., Monk & Osborne, 1997), the most compelling single summative assessment study was the complex, multifaceted, randomized research design for assessing Harvard Project Physics (HPP), reviewed by Welch in 1973. He stated that compared with their counterparts in non-HPP classes, "students in HPP find the course more satisfying, diverse, historical, philosophical, humanitarian, and social" (p. 375). Welch also reported that standardized measures of humanistic conceptual content achievement (i.e., pre-posttest gain scores on Klopfer's "Test on Understanding Science" and Welch's "Science Process Inventory") showed no significant difference between the HPP and non-HPP groups. However, I discovered (Aikenhead, 1974) that Welch (1973) and Welch and Walberg (1972) had reached their conclusions on the basis of compromised and skewed data. About 4% of the data comprised frivolous responses by uncooperative students (e.g., students penciled in the response boxes on the machine-scored answer sheet in a way that spelled out an obscene expletive suggesting a physical impossibility). About 12% of the data came from grossly incomplete responses that dramatically and spuriously skewed the pre-posttest gain scores. By deleting the frivolous and incomplete answer sheets, I recalculated the gain scores and found the HPP group had significantly outperformed the non-HPP group.

I posited two conclusions about quantitative research (Aikenhead, 1974). First, researchers need to take the time to ensure they have "clean" data to analyze. This is particularly true today for data collected digitally. Second, "What does it mean to a curriculum developer or teacher for group E [experimental] to score 3.77 points more than group C [control]?" (p. 23). Such ambiguous summative data have motivated many researchers to take up the interpretive paradigm.

A note of caution: The deficit model of assessment continues to hold sway in most quasi-experimental studies, especially in the political arenas of curriculum policy and international summative assessments, even though the model can create false crises (Gibbs & Fox, 1999).

The Third International Mathematics and Science Study (TIMSS) included in its quantitative data collection a cluster of humanistic themes around the history, philosophy, and sociology of science (Wang & Schmidt, 2001). Using data tabulated from several domains within TIMSS (e.g., curriculum analysis, textbook analysis, teacher questionnaire, and students' scores on canonical science content), the researchers concluded that the engagement of students in humanistic content was significantly associated with their general school science performance, in the handful of countries that provided such a curriculum. Items addressing humanistic content itself were all but absent in TIMSS achievement tests. Fensham (1994) suggested two reasons for this: They are very difficult to compose because of their contextualized nature; and they generally have low status among the 40 participating countries.

OTHER INVESTIGATIONS IN HUMANISTIC SCHOOL SCIENCE

Many useful findings have accumulated in the research literature to support and guide the expansion of humanistic school science worldwide. With the development of each new evidence-gathering technique for humanistic science content, a status report has often been published to provide baseline data for a general sample of students. These results have posited politically useful evidence (Leach et al., 1997), particularly when a humanistic perspective is being introduced into a country (Manassero-Mas & Vázquez-Alonso, 1998).

For instance, Driver et al. (1996, pp. 131–134) were able to conclude with high confidence that United Kingdom students conceived of science as a social enterprise but in dangerously simplistic ways. The following simplistic notions were expressed: Disagreements between scientists were quickly explained by students as biases of scientists or a lack of facts (but students ignored, or did not know about, scientists' different but legitimate value positions or their different conceptual perspectives). Students expected scientists to produce unambiguous and incontrovertible facts, that is, conclusive evidence and not circumstantial evidence. Students showed little awareness of the internal and external social factors at play in the development and extension of scientific knowledge. The United Kingdom students did, however, have a rich enough prerequisite knowledge of human nature and social institutions to apply that knowledge rationally to scientific controversies, but students needed the guidance of a humanistic science teacher ("explicit curricular interventions," p. 134) to help them apply their prerequisite general knowledge instead of their simplistic preconceptions. Similar types of conclusions were reached by other status studies using different instrumentation (e.g., Lederman et al., 2002; Ryan & Aikenhead, 1992).

By analyzing answers to a national examination for an STS syllabus in the United Kingdom in the 1980s, Solomon (1988a) concluded that explicit instruction had made a difference to students' capability to construct plausible arguments from opposing points of view on a science-related topic. About a decade later a new humanistic syllabus appeared, Science for Public Understanding (Millar, 2000), and its impact on students was assessed by Osborne, Duschl, et al. (2003). This integrated science course was designed for postcompulsory students who wanted to broaden their understanding of science. The 78 teachers who participated in the assessment study had exemplary credentials, by and large, and taught from a textbook especially developed for the syllabus (Hunt & Millar, 2000). The research data came from questionnaires, examination papers, classroom observations, and interviews with teachers and with students. Science for Public Understanding was highly successful at attracting nonscience students into the course, which they found interesting, enjoyable, and equally difficult (according to examination results of their other courses). This was a major achievement for a senior high school level science course, according to the researchers, who were particularly impressed by the students' high degree of interactivity and their authentic engagement with various aspects of the course.

Student interest was also cited as a major achievement in Häussler and Hoffmann's (2000) humanistic physics curriculum in Germany (primarily functional science; Chapter 3). Graduates of the course perceived physics "more as a human enterprise and less as a body of knowledge and procedures" (p. 704). Students also valued the curriculum's cultural relevance, as indicated by their increased self-esteem, a fundamental outcome to any learned curriculum.

Students' ability to interpret the news media is another expectation of most humanistic curricula (Fensham, 2000a, 2000b; Stocklmayer et al., 2001; Thier & Nagle, 1994). Ratcliffe (1999), for instance, investigated the evaluation reports written by three groups of students critiquing science articles in the *New Scientist*. The groups consisted of school-aged students, college science students, and science graduates. Although students' skills increased with formal training, years of experience, and self-selection into science, as one would expect, Ratcliffe discovered that the skills of evidence evaluation (based on concepts of evidence, a component of functional science; Chapter 3) were evident across all three populations, and she suggested that these abilities could be developed further in school science through explicit teaching.

The impact of history of science materials on student learning was investigated by Solomon et al. (1992) in an elaborate 18-month action research project. Interestingly, students' facile, media-icon images of scientists were not *replaced by* realistic images developed through learning the humanistic content, but instead these realistic images were *added to* the preconceptions in students' minds (i.e., concept proliferation rather than concept replacement). From a student's point of view, learning meant they now had a choice between two images, and the choice depended on context. This result has implications for the importance of context in the assessment of student learning. Solomon and her colleagues' evidence also suggested that learning scientific theories was more durable for students who had learned the reasons for accepting one theory over another. The stories from the history of science smoothed the path for students' conceptual change.

In a follow-up study, more focused on seven images of scientists and their experiments, Solomon et al. (1994) confirmed earlier results and further concluded: "The stories of the actual activities of scientists are memorable enough to create a valuable library of epistemological ideas" (p. 372), but the stories do not erase simplistic icon images of scientists—a result that was also found in Lin's (1998) research with preservice teachers (Chapter 5). Extending their research program into a large-scale study of 1,000 students, Solomon et al. (1996) investigated the possible interaction between two domains of knowledge: ideas gained through one's own activities in school science, and ideas held about the authentic activities of professional scientists. The researchers concluded that each domain was predisposed to different sorts of learning. The researchers also discovered that students were unfamiliar with scientific theories per se, and that students' cartoonlike images of scientists developed into more authentic or realistic images.

As stated at the beginning of this chapter, socially responsible action by students is a valued outcome of the learned curriculum for many humanistic science curricula (Cross & Price, 1992, 2002; Rubba, 1987; Solomon, 1994b, 1994d). Accordingly, Solomon (1990, 1992) initiated a 3-year study in which teachers infused humanistic content purposefully throughout their science courses by using news clips from television programs to initiate open, informal, small-group, student discussions on emotion-laden science-related topics (e.g., incidence of leukemia), with little guidance from the teacher. By analyzing the transcripts of these discussions and pre/post questionnaires, Solomon (1990) discovered, along with Fleming (1986b) and Levinson (2004), that merely a simple familiarity with scientific terms used in the news clips (i.e., enticed-to-know science) made participation in the discussions easier and more effective, as did familiarity with civic knowledge and moral reasoning (Solomon, 1994b). Solomon also concluded that students tended to become more cognizant of their civic responsibility in terms of being self-reliant in making up their own minds on an issue. But when analyzing the 3 years of student transcripts, she was unable to detect a pattern of association between student knowledge or attitudes and students' actual behaviors as recorded in the transcripts, a finding at odds with some modestly positive assessments of student action resulting from specific humanistic science experiences (e.g., Wiesenmayer & Rubba, 1999).

The relationship between student knowledge/attitude and responsible action is fraught with complexity (Rickinson, 2001). Quantitative syntheses of research described this relationship as a meager correlation (Posch, 1993), or a small effect size of 0.3 (Hines, Hungerford, & Tomera, 1987). Qualitative studies have pointed to important influences on action (Dahncke, 1996; Kortland, 1992; Solomon, 1994d): social pressure, economic constraints, cultural factors, and students' belief in their capacity to bring about change (i.e., potential political potency). However, the research synthesized in this chapter suggests that socially responsible action can be enhanced by humanistic school science for some students with certain teachers. A major component to this responsibility is one's willingness to engage in thoughtful decision making.

STUDENT DECISION MAKING

The wise use of knowledge, scientific or otherwise, enables people to assume social responsibilities expected of attentive citizens or key decision makers employed in public service, business, and industry. Thus decision making is often central to a humanistic science curriculum, and it serves as a vehicle to transport students into their everyday world of need-to-know science, functional science, enticed-to-know science, have-cause-to-know science, personal-curiosity science, and science-as-culture (Chapter 3).

A typical lesson creates a simulation of an everyday event, the type that led researchers to conclude (Chapter 3) that when people need to communicate with experts and/or take action, they usually learn the science content required and that this content will often be action-oriented science (citizen science)—that is, interdisciplinary canonical science deconstructed and then reconstructed to fit the unique circumstances of the everyday event (Jenkins, 1992; Lawrence & Eisenhart, 2002; Layton, 1991).

But decision making necessarily encompasses other types of knowledge: always values and personal knowledge, and sometimes technology, ethics, civics, politics, the law, economics, public policy, and so on (Aikenhead, 1980; Dawson & Taylor, 2000; Driver et al., 2000; Grace & Ratcliffe, 2002; Jiménez-Aleizandre & Pereiro-Muñoz, 2002; Kolstø, 2001a; Oulton, Dillon, & Grace, 2004; Thomas, 2000; Zeidler, 2003). For instance, in research into the conflicting testimonies of scientific experts on science-related events and issues, even the scientific technical information itself was found to carry political-ideological baggage (i.e., values); and rather than achieving a clear resolution on an issue, more scientific information invariably caused greater polarization (Aikenhead, 1985; Gaskell, 1982). Fensham (2002) argued, "The reason that different groups of scientists can often differ in their assessment of such issues is not so much that one group is right and the other wrong; rather it is that both are right, but about different aspects of the issue . . . [depending] on the wider value positions of the groups themselves" (p. 16).

How do students make a decision in the context of conflicting expert advice from scientists? The research question has been investigated epistemologically, sociologically, and axiologically. The accumulated evidence clearly advises that the following humanistic objectives need to be achieved by students in order to communicate effectively with others about a socioscientific event or issue: to learn to ask pertinent questions, to obtain evidence, and to use it as the basis for decision making; to understand the characteristics and limitations of scientific evidence; to understand the methodological norms in science; to understand the social processes engaged in by scientists; to understand the institutional aspects of science; to identify the value positions or ideologies held by scientists on each side of a debate; and to have access to appropriate social criteria for judging credibility of scientists (Duschl & Gitomer, 1996; Gaskell, 1994; Kolstø, 2000, 2001a, 2001b; Kolstø et al., 2004). Trust in scientific information is determined more by students' identifying value positions related to that information than by their understanding the specific information (i.e., canonical science content; Fensham, 2002). In short, humanistic content (i.e., knowledge *about* science and scientists) is more relevant than canonical science content in the context of conflicting expert advice (Aikenhead, 1985; Bingle & Gaskell, 1994; Ryder, 2001). Disagreements among expert scientists make activities more authentic but at the same time more difficult for students (Kolstø, 2001a; Solomon, 1988b). Linn, Clark, and Slotta (2003) have designed a learning environment to assist students by scaffolding their interactions with science-related issues.

Other avenues of research into student learning have been pursued in the context of decision making. Outside the milieu of school science, Fleming (1986a, 1986b) meticulously examined how 17-year-old students individually reached decisions on proposed socioscientific issues, for example, accepting employment in a nuclear power plant. Using multistage interviewing techniques, he came to the conclusion that students made their decisions primarily by reasoning in the domains of either "moral issues" or "personal reasoning" (reasoning devoted to the maintenance of the self) rather than by evidence-based reasoning endemic to scientific decision making (Duschl & Gitomer, 1996). (Adults sometimes reason in the domain of "social conventions," but no evidence of this surfaced in Fleming's student data.) Fleming's participants ignored relevant scientific information offered to them because they perceived scientists as being interested only in progress unrelated to human welfare. In general, students seem to compartmentalize scientific information and isolate it from information used when they make personal decisions (Sadler, 2004). Research into cross-cultural science adds weight to these results; most students' worldviews of nature are dramatically different from those of their science teachers (Cobern & Aikenhead, 1998; Cobern et al., 1999), suggesting a more fundamental reason (e.g., cultural self-identity) for the divide between Fleming's students' decision making and scientific evidence-based decision making. Curriculum policy makers, of course, had assumed evidence-based decision making should be evident in students' decisions.

Fleming's research seriously questioned the importance of using scientific knowledge when making a decision on a socioscientific issue in the everyday world, a finding replicated many times in the literature since then (e.g., Layton et al., 1993; Solomon, 1988b, 1992; Tytler et al., 2001a). His research encouraged science educators to design relevant decision-making events for the classroom in which students would learn scientific ideas and learn how to use those ideas appropriately rather than ignore them completely.

Many researchers rose to the challenge. Because of space limitations, my synthesis of their findings omits research details that speak to the care and depth of their investigations.

Decision-making models from sociology and psychology failed to account for the complexities of classroom decisions associated with socioscientific events and issues (Aikenhead, 1989). Thus other models needed to be developed for science education. Ratcliffe (1997b) distinguished between descriptive models (what students actually do) and normative models of decision making (what students should do). Each type has guided different research studies.

Descriptive models were sought in studies that investigated how students made decisions after discussing a science-related event, but without the benefit of specific instructions on how to make a good decision (Eijkelhof, 1990; Kolstø, 2001a; Ratcliffe, 1997b). By discovering students' untutored ways of coming to a decision, researchers clarified the challenges that lay ahead for teachers and curriculum developers who plan to implement a humanistic approach. Decision making in classrooms was found to be much more complex than originally assumed. Today we have a more detailed and realistic understanding of socioscientific decision making. For instance, Ratcliffe (1997b) developed a descriptive decision-making model for small-group work. In addition to her interest in students' use of scientific ideas (students seldom referred to science content in her study), Ratcliffe's research looked at how students conducted their decision making and what values guided them. She empirically identified the following list of key features of successful small-group decisions: "understanding procedures for rational analysis of the problem; awareness and use of available information; clarification of the concerns and values raised by the issue; recognition of how scientific evidence may assist in the decision; motivation to engage fully in discussing the issue; and consideration of and respect for differing viewpoints about the issue" (p. 167). Students were guided by different sets of values depending upon the socioscientific issue. Ratcliffe observed in students "an ability to identify suitable options" and "an ability to identify criteria at some stage in the discussion but difficulty in using them systematically in reasoning" (p. 175).

In a much different study, Kolstø (2001a) discovered the following five descriptive models depicting how individual students made a decision on whether or not to bury high-voltage power lines in a populated area: relative risk model, play safe model, uncertainty model, small risk model, and a pros-and-cons model. All five entailed values. When interviewed afterward, only a small minority of students stated that canonical content (e.g., voltage and electromagnetic field fluctuations) was desirable but not necessary, whereas a large majority found canonical science content neither desirable nor necessary, a conclusion replicated in other studies (Eijkelhof, 1990; Grace & Ratcliffe, 2002; Kortland, 2001).

One notable exception to this result allows us to identify other complexities in decision making. In an extensive study that involved grade 11 Spanish students in authentic community-based activities, Jiménez-Aleizandre and Pereiro-Muñoz (2002) explored students' scientific knowledge and argumentation skills required to reach socioscientific decisions on environmental management, and compared these skills to those of scientific experts. The researchers elaborated Kortland's (1996) emphasis on argumentation, drew on Ratcliffe's (1997b) decision-making model, and followed Duschl and Gitomer's (1996) notion of authentic problem solving. The students' success at applying canonical science and at approximating expert arguments was attributed to the students' becoming "a knowledge-production community" (as opposed to more passive consumers of knowledge), a role that judiciously combined values with scientific ideas and data. The authenticity achieved in this study approximated that from the case-study research into citizens' acting upon a science-related issue (Chapter 3). On the one hand, the Spanish study exemplifies the conclusion repeated earlier that when people believe they need to take action, they tend to learn the science required (often science-as-culture). On the other hand, a unique feature of this study compromises its transferability to other classrooms: The class was taught by one of the researchers whose orientation to humanistic science would likely be unusually strong. This makes the results less transferable to classrooms taught by regular humanistic science teachers or middle-of-the-road science teachers. Nevertheless, Jiménez-Aleizandre and Pereiro-Muñoz's (2002) excellent research design sets a high standard for others to follow.

Values are a constant feature of decision making. They include economic, ethical, religious, social, and political values. Researchers have consistently found that students give much higher priority to values, common sense, and personal experience than to scientific knowledge and evidence, even when that knowledge and evidence are relevant to the issue discussed in the classroom (Bell & Lederman, 2003; Dawson & Taylor, 2000; Grace & Ratcliffe, 2002; Kolstø, 2001a; Levinson, 2004; Ratcliffe, 1997b; Sadler, 2004).

Perhaps the scientific concepts taught in those studies were chosen by researchers or teachers to represent the curriculum's canonical content whose scientific validity was highly secure, what Cole (1992) called "core science" (e.g., electromagnetic waves), whereas the scientific concepts critical to resolving a socioscientific issue from a student's point of view may have had a much more tentative validity, "frontier science" (Cole, 1992) (e.g., the possible link between high-voltage power lines and childhood leukemia). Scientists themselves will disagree over the validity of frontier science, creating more uncertainty, and thus by default giving students more reason to rely on their own values when making a socioscientific decision in the context of uncertain science. Quoted above, Fensham (2002) recognized that salient features of a socioscientific issue vary according to a scientist's differing value positions (ideologies), thereby causing disagreement among scientists and, hence, creating controversy.

Fundamental scientific rules and criteria of normal science (Kuhn, 1962) do not apply to the high risk and high uncertainty associated with socio-environmental systems, which require rules and criteria of "postnormal" science (Ravetz, 1997). The greater the reliance on frontier science, the greater the controversy will be. When there is less certainty in the knowledge (e.g., frontier science), there is a greater need to determine the credibility or trustworthiness of the sources of scientific information and, consequently, there is a greater reliance on values. This conclusion helps to explain the apparent low status, or nonexistence, of canonical science (core science) in decision making on most science-related controversial issues.

Similarly, research into students' understanding of the nature of science and its effect on student decision making indicates that students' values and personal experiences are more influential than their understanding of the nature of science, realistic or naïve (Bell & Lederman, 2003; Sadler, 2004; Zeidler, Walker, Ackett, & Simmons, 2002).

In addition, we need to consider the emotional tenor of an issue. Emotionally charged issues tend to cause students to rely more on their personal values (Grace & Ratcliffe, 2002; Levinson, 2004). However, when students feel a personal and responsible attachment to an issue (e.g., due to the prospect of participating in an action as a consequence to their decision), they seem to be more prone to consider scientific ideas and evidence (Jiménez-Aleizandre & Pereiro-Muñoz, 2002; Roth & Désautels, 2004). In science classrooms, the inclusion of science-related decision making about either artificial or real-life events does not necessarily promote the motivation to use science content (Sadler, 2004). Instead, anticipated action seems to capture authenticity for students, for instance, a personalized local issue (Patronis, Potrari, & Spiliotopoulou, 1999; Pedretti, 2004; Watts et al., 1997). This science content for taking action may be a deconstructed and reconstructed version of canonical science content for explaining phenomena (Chapter 3).

Another problem to consider concerns students' conceptions of scientific evidence, that is, their concepts of evidence (Duggan & Gott, 2002), a component of functional science described in Chapter 3. Without overt instruction, teachers can anticipate that about one-half of their students will hold naïve conceptions about scientific data (Jiménez-Aleizandre, Agraso, & Eirexas, 2004; Sadler, Chambers, & Zeidler, 2004; Tytler et al., 2001a), and therefore scientific evidence will be either misused or ignored in favor of common sense and personal experience (Sadler, 2004), and thoughtful decision making will suffer as a result. Humanistic science teaching needs to give special attention to teaching concepts of evidence.

For some students, overt instruction on how to make a thoughtful decision (i.e., metacognition concerning normative decision-making models) made a difference to student learning (Eijkelhof, 1990; Kortland, 1992, 1996, 2001). Ratcliffe (1997b) also found that students who already possessed reasonable decision-making skills improved upon these skills through group argumentation, but students with low-level skills did not. Kortland's (1992) research project structured the decision-making process in particular ways for students. The process was guided by his normative decision-making model. Kortland assessed students' decision-making skills by the quality of their arguments, particularly the range, depth, and explicit weighing of their arguments. However, students were held back more by a lack of content knowledge about garbage management (i.e., functional science) than by deficiencies in their decision-making skills. Similar results for genetic engineering and for DNA testing were reported by Sadler and Zeidler (2004) and Dawson (2004), respectively. On the basis of Kortland's (1996) evaluation data, he constructed a four-tiered, developmental decision-making scheme to guide humanistic science educators in their own development of curriculum materials aimed at enhancing students' decision-making skills (i.e., their ability to present an argued point of view). Upon further research, he discovered the interplay between group decisions and individual decisions (Kortland, 2001), and noted an improvement on some aspects of argumentation but not on others, results similar to those of Zohar and Nemet (2001) and Dawson and Taylor (2000). Other normative decision-making models have guided research into student decision making (Aikenhead, 1985, 1994a; Cross & Price, 2002; Dahncke, 1996).

In addition to noting the effectiveness of his decision-making model, Kortland (2001) identified an important methodological finding: Students' oral or written presentations of their argued point of view often failed to communicate the quality of thinking that had actually gone into their decision. Zeidler's (1997) and Driver et al.'s (2000) research into student difficulties with argumentation adds a cautionary practical note for humanistic science educators who have overly high expectations of students' capabilities at argumentation, a topic of continuing research today (Jiménez-Aleizandre et al., 2004; Sadler, 2004).

Some researchers have focused on students' success at systematic ethical or moral reasoning in science-related ethical issues (Zeidler, 2003). When evaluating a module about DNA testing, for instance, Dawson (2004) discovered that some students learned to correctly identify ethical issues that such testing raises. Ratcliffe's (1997a) positive results with her participants led Reiss (1999) to conclude that science teachers who lack specialized knowledge in moral philosophy can still assess students' ability to reason ethically on socioscientific issues. For example, Solomon's (1994b) categories of moral statements proved to be very useful to Pedretti's (1999) research into assessing student ethical decision making at the Ontario Science Centre.

Concern over the ethical maturity of 14- and 15-year-olds surfaced in Dawson and Taylor's (2000) study when students' bioethical decision making was found to be naïve, narrow, and idealistic, even though the students had been taught four key bioethical principles and a normative decision-making model. This failure could have been predicted, however, from Driver et al.'s (1996) major study in which they discovered that students generally need to be at least 16 years of age to make thoughtful decisions on controversial issues (Chapter 1). Ratcliffe (1997b) reported that 15-year-olds were able *to begin* to participate in thoughtful decision making on ecological issues.

The research on decision making synthesized here generally consisted of preliminary studies that revealed multiple influences on student decision making (e.g., age of student, overt instruction, authenticity of the activity, and emotional content) interacting simultaneously with various predispositions of students. By knowing students' preinstructional ways of making a socioscientific decision, humanistic curriculum developers and teachers now have clearer ideas of where to begin. The research also uncovered useful distinctions between, for example, core science and frontier science (Cole, 1992). Such distinctions must become standard humanistic content for thoughtful decision makers; otherwise cynicism toward all of science may result from students' misunderstanding frontier science as being as authoritative as core science, or vice versa (Thomas, 2000).

The research suggests that the degree to which scientific concepts are used by some students in a socioscientific decision is predicated on three factors. First, the actual relevance of the concept needs to be established. Second, students must have a meaningful understanding of the concepts. Third, when the first two factors are fulfilled, students still need to perceive the connection between the concept and decision issue. These three conditions are rarely met, however.

Perhaps the most pervasive result from the research into student decision making was the priority students gave to values over scientific ideas and evidence. This result may be due to the fact that values *are* more important for deciding most science-related events and issues, even for science teachers and scientists themselves (Law, 2002). Lawrenz and Gray (1995) discovered that science teachers with science degrees did not use science content to make meaning out of an everyday event but instead used other knowledge such as values. Bell and Lederman (2003) investigated how 21 university research scientists made decisions on socio-scientific issues (e.g., fetal tissue implantations, global warming, and smoking and cancer). Using questionnaires and telephone interviews, the researchers concluded that all participants considered the scientific evidence, but they "based their decisions primarily on personal values, morals/ethics, and social concerns" (p. 352). Should we expect students to be any different?

CONCLUSION

The research literature unequivocally demonstrates that student learning (defined by various objectives) does occur to varying degrees as a result of a humanistic science curriculum enacted by competent and supported teachers. Over the years, our early naïvety about this learning has been replaced by a more realistic complex of paradoxes and trade-offs. Relevant everyday contexts alone did not necessarily nurture greater canonical science attainment, although nothing of significance was lost on this count either. Values and student experiences played a large role in focusing students' attention on both humanistic and canonical science content: The more emotional the context of instruction or the more uncertain the relevant scientific information was, the more important values became, and thus the attention paid to canonical science content lessened. Students tended to learn more humanistic content the more explicit it was in their classroom instruction and assessment. At the same time, studies also showed that the more overt this humanistic content became, the greater disruption it caused to status quo teaching, and thus the greater challenge it posed to teachers. Moreover, when we apply constructivist principles to students' learning humanistic content, the challenge is even more difficult, especially for students who simply want to memorize content (i.e., play Fatima's rules).

One ubiquitous research result from many studies into student learning was the positive reaction of most students. As stated in Chapter 5 but well worth repeating here, Osborne, Duschl, et al. (2003) articulated a key implication to their research in the U.K. Science for Public Understanding syllabus: "Teaching a course which is enjoyed by students is . . . much more engaging and motivating for the science teacher" (p. 19). Häussler and Hoffmann (2000) found that students' positive reactions coincided with stronger self-esteem as a result of being successful achievers. More studies should collect data on students' self-esteem and self-identity, because the two represent fundamental outcomes of a humanistic science curriculum (Chapter 7). Ramsden (1992) heard a slightly different response from students who had enjoyed a humanistic science curriculum: They seriously questioned whether it was proper school science.

In general, the positive reaction of most students to humanistic school science was likely due to a number of factors: a genuine interest on the part of students, a happy diversion from teacher-centered instruction, and a selection bias in the research sample favoring the large majority of students who generally find the traditional curriculum boring and irrelevant. This last point implies a cautionary note that the small minority of science-proficient students who embrace a pipeline ideology for school science will likely not respond positively to a humanistic perspective in their science courses (Chapter 5). Similarly, students who politically equate school science with future earnings, even though school science holds no intrinsic value for them, will likely resist a humanistic perspective as well (Désautels et al., 2002).

7

Culture Studies: School Science as Culture Transmission

Some researchers in science education have discovered that cultural anthropology can be fruitful to their work for several reasons: It highlights a humanistic perspective; it offers new avenues of inquiry; and it promises insightful descriptions and explanations to benefit policy makers, teacher educators, and classroom teachers. Culture studies in science education is a new research field within anthropology of education. It views teaching and learning as culture transmission and acquisition, respectively (Chapter 2), not as playing Fatima's rules (Chapter 3).

Humanistic perspectives in science education for grades 6 to 12 challenge the dominant pipeline ideology that seeks to enculturate all students into scientific disciplines by means of transmitting the culture of science to them (Chapter 2). Humanistic school science provides alternative cultural transmissions for students. This chapter explores the research into these alternatives.

CROSS-CULTURAL SCIENCE EDUCATION

Culture-based clashes occur in science classrooms for students whose worldviews and cultures (including their home language) differ from those of Western science conveyed by school science (Allen & Crawley, 1998; Baker & Taylor, 1995; Brown, 2004; Cajete, 2000b; Calabrese Barton, 2001b; Chang & Rosiek, 2003; Cobern, 2000; Contreras & Lee, 1990; George, 1999b; Jegede, 1995; Kawagley, 1995; Kozoll & Osborne, 2004; Krugly-Smolska, 1995; Lee, 1997; MacIvor, 1995; Maddock, 1981; McKinley et al., 1992; Norman et al., 2001; Ogawa, 1995; Ogunniyi,1988; Okebukola & Jegede, 1990; Rowland & Adkins, 2003; Rollnick & Rutherford, 1996; Tharp, 1989; Tsai, 2001b). This happens in spite of supportive influences on student learning. Discordant worldviews create an incompatibility between, on the one hand, students' self-identities (e.g., who they are, where they have been, where they are going, and who they want to become; Cajete, 2000a) and, on the other hand:

- students' views of Western science, school science, or their science teacher (Aikenhead, 1996; Atwater & Crockett, 2003; Brickhouse et al., 2000; Cobern & Aikenhead, 1998; Eisenhart et al., 1996; Gilbert & Yerrick, 2001; Kawagley et al., 1998; McKinley et al., 2004; Tobin, 1997), and
- students' views of the kind of person they think they must become in order to engage in science (Brown, 2004; Calabrese Barton, 1998; Carlone, 2004; Gagné, 2003).

Students who do not feel comfortable taking on a school science identity (i.e., being able to talk, think, and believe like a scientist) represent the vast majority of any student population (Atkin & Helms, 1993; Brown et al., 2004; Cobern et al., 1999; Costa, 1995; Eisenhart et al., 1996; Reiss, 2000).

The evolution of a humanistic perspective in school science over the past 50 years first targeted Euro-American students and their enculturation into Western society (Chapter 2). When students in non–Euro-American nations became a concern to researchers (Maddock, 1981), a new question arose: Whose culture is being transmitted in school science? If Western ways of knowing and valuing continue to dominate science teaching for these students, then science classrooms continue to be instruments of colonization because Western cultural ways of knowing are transmitted with the expectation that students will reject their indigenous ways of knowing to participate in the classroom's community of science learners. This cultural assimilation has been advocated by some researchers with allegiances to the ideology of preprofessional training (e.g., Lubben & Campbell, 1996; Lynch, 1996).

The severity of culture clash for these students tends to increase the more a student's home culture differs from the culture of Western science portrayed in schools, for instance, in the cases of Aboriginal students in Western nations and worldwide (Aikenhead, 1997; Battiste & Barman, 1995; Cajete, 1999; Chinn, 1999; Ezeife, 2003; George, 1999a; Jegede & Okebukola, 1991; Kawagley, 1990; McKinley et al., 1992; Sutherland & Dennick, 2002). This is also the case to some extent for Asian students (Cobern, 1996a; Kawasaki, 1996, 2002; Lee, 1997; Ogawa, 1995; Tsai, 2001b), but not all Asian students (Koul, 2003). When students' cultural self-identities do not harmonize with the cultures of scientific disciplines, many students enter a science classroom feeling as if they have crossed a type of cultural border. For them, *learning school science is a cross-cultural event.*

The field of culture studies in science education is attempting to clarify the cultural transmission appropriate for these students. The field is complex owing to the heterogeneity of cultural contexts of schools; students' worldviews and self-identities; and Western culture itself, in which science and technology constitute only a subculture, albeit a very powerful subculture, among diverse Western subcultures. Heterogeneity is illustrated in Costa's (1995) study by two students from the same family who have quite different reactions to high school science. Because culture studies in science education is an emerging research field, its findings are limited at this time.

In any non-Western culture there is always a minority of students whose worldviews harmonize with a worldview endemic to Western science. These students want to be enculturated into Western science to enhance their self-identities. Many drop out of the pipeline (Chapter 3), but some persevere in spite of obstacles, and these students become an inspiration to other like-minded students. Case studies of these students abound (e.g., Becher, 2002; Brandt, 2004; Chinn, 1999; McKinley, 2003).

At the other extreme are students (mentioned earlier) with esthetic or spiritual worldviews who, for instance, avoid culture clashes by resisting school science (e.g., by choosing not to enroll in science courses or by playing Fatima's rules if required to take a course; Chapter 3).

This chapter conveys the research community's response to this full spectrum of non-Western students (from the science-proficient to the science-phobic) for whom the disciplines of Western science are in some way foreign cultures. By attending to the cases of some non-Western students in science classrooms, the chapter constructs practical concepts that are transferable to a large majority of Western students in most science classrooms, the target population of previous chapters. (For a specific Islamic perspective, see Irzik [1998] and Loo [2001].)

The field of culture studies in science education ultimately applies to all students who experience school science as a cross-cultural event. This excludes, of course, the privileged elite who find the science pipeline highly comfortable as it is. (As a student, I belonged to that group.)

A number of symposia and special issues in journals have brought together researchers in cross-cultural, multicultural, and urban science education to clarify and advance emergent research programs. This literature is synthesized here according to the following topics: guidelines for respecting students' cultures, indigenous sciences, student heterogeneity, and issues related to cross-cultural (bicultural) teaching and learning.

GUIDELINES FOR RESPECT

In the political reality of knowledge–power relationships elucidated by postcolonial thinking (Battiste, 2000; Battiste & Barman, 1995; Bishop & Glynn, 1999; Smith, 1999), science education researchers have conventionally occupied the class of privileged colonizer. Guidelines for respecting students' cultures can help us confront our Eurocentric upbringing and help us disrupt the historical marginalization of Aboriginal peoples. This process of confronting and disrupting will sharpen our vigilance against inadvertent discrimination or colonization on our part. Drawing upon postcolonial research in science education (Cajete, 1999, 2000a, 2000b; George, 1999b; Kawagley et al., 1998; Kawasaki, 2002; Keane & Malcolm, 2003; MacIvor, 1995; McKinley, 2003; McKinley et al., 2004; Ninnes, 2000, 2004) and upon the wisdom of elders who contributed to the *Rekindling Traditions* project (Aikenhead, 2000a, Teacher Guide) and to the Alaska Native Knowledge Network (ANKN, 2004a), the following guidelines are offered. All apply to discussions of Aboriginal students found in this book, and many apply to cross-cultural school science for non-Aboriginal students.

Every individual is a dynamic unique hybrid of many ongoing life experiences and genetic predispositions. This uniqueness is truncated whenever we identify individuals as members of a group or culture. Grouping is unavoidable, however, in our nonideal world of education if we are to make headway improving science education for marginalized students. By embracing the complexities of heterogeneity, we can avoid stereotyping students and cultures (i.e., "essentializing" them; Erickson, 2004). Similarly our Eurocentric habit of using dichotomies (e.g., marginalized versus mainstream) can in some contexts create an asymmetric power relationship that privileges our Eurocentric status. (Nevertheless, dichotomies continue to be useful as discursive devices to define a continuum.) The guideline here is to be sensitive to our use of categories by not stereotyping group members.

Labels themselves can unconsciously withdraw or confer status. For example, the label *ethnobotany* applied only to a field of indigenous science obliterates the culture-laden character of university botany (Euro-ethnobotany), thereby privileging the academy's science.

Translate with caution. The act of translating Western science into indigenous languages, and vice versa, is fraught with problems of incommensurability. Translations lose epistemological, ontological, and axiological nuances, and hence cause misunderstandings.

We should avoid prescribing authenticity to events or to identities within a group but instead accept the judgment of indigenous spokespersons for that group when they describe what is authentically traditional for them *today*. We can describe a traditional indigenous science in the present verb tense rather than relegate it to a distant millennium by describing it in the past tense. Aboriginal cultures evolve; they are dynamic; they acculturate European technologies; and thus traditional indigenous sciences change.

When we include indigenous science in school science, we should avoid tokenism. A community's indigenous perspective and language could serve as the framework for teaching whenever possible, not as an appendage to a unit.

Authenticity is place-based. An indigenous science has meaning only when connected to the community's land. Indigenous science is not universal. We must embrace the natural diversity of knowledge held among indigenous communities. By citing contradictions, we merely confirm that indigenous science is not Western science. A sense of place is also endemic to urban science education (Calabrese Barton, 2002, 2004).

Knowledge acquired from most indigenous peoples represents a relationship between the researcher and the person who shared that knowledge. Indigenous science should not be isolated from this relationship. We acknowledge the relationship when we mention the permission we received to share it with others.

Learning indigenous science is a journey toward wisdom, not an accumulation of facts and concepts. The English word *knowledge* has no clear counterpart in most Aboriginal languages. The best translation is "ways of living." This has enormous implications for assessing Aboriginal students.

We should write about indigenous science (ways of living) in the context of holism, not Cartesian duality and reductionism. An indigenous fact about nature is as much about values and spirituality as it is about a natural phenomenon.

Indigenous science (ways of living) is multigenerational. Think of curriculum content as being shared with the seventh generation yet unborn.

We can privilege marginalized groups by articulating and acknowledging the challenges they often face when they deal with us ("the other"), who represent the dominant colonizing culture. Our language ought to reflect ourselves as "the other."

The purpose of science education should be established by the school's community and not by Eurocentric science educators. Although there is no consistent agreement within indigenous communities concerning this purpose, a consensus has formed around cross-cultural (bicultural) approaches to science education. "Creating a balance between two worldviews is the great challenge facing modern educators" (Battiste, 2000, p. 202). Many indigenous communities stipulate that students need to learn to live well in both worlds, indigenous and Western, in ways that respect the personal talents (gifts) and cultural self-identities of students. This entails making Western science accessible to students for the well-being of their culture.

The few Aboriginal students who do have a Western sciencelike worldview should receive a culturally responsive enculturation into the culture of Western science. They are the future engineers, teachers, scientists, and health professionals needed for the economic development, ecological sustainability, self-governance, and cultural survival of their communities. An ideal goal is to have these students "master and critique scientific ways of knowing without, in the process, sacrificing their own personally and culturally constructed ways of knowing" (O'Loughlin, 1992, p. 791).

On the other hand, students who possess other worldviews should receive an enculturation into their own culture to become savvy citizens who deal effectively with hegemonic Western science and technology (Chapters 3 to 6); they too are responsible for the survival of their culture. Students can learn to use Western science as a tool without becoming identified with that tool, an achievement particularly celebrated by Japanese science educators (Ogawa, 1997; Sumida & Kawasaki, 2004).

Humanistic school science nurtures both types of enculturation for Aboriginal students. This goal is also transferable to other marginalized groups of students who feel that school science is a foreign culture. For example, Tobin, Seiler, et al. (1999) documented in detail how a school, by not respecting the local culture of an inner-city community, enhanced students' resistance to school science, in spite of the teacher's efforts to the contrary.

INDIGENOUS SCIENCES

To achieve the cross-cultural (bicultural) school science envisaged in many communities, indigenous and Western sciences should be integrated in the school's curriculum and instruction (Cajete, 1999; Kawagley, 1995; MacIvor, 1995; McKinley et al., 1992). Similarities and differences between the two types of knowledge systems have received much attention (e.g., Aikenhead, 1997, 2001a; Cajete, 2000b; Christie, 1991; George, 1999a; Hammond & Brandt, 2004; Jegede, 1995; Kawagley et al., 1998; McKinley et al., 1992; Snively & Corsiglia, 2001). These similarities and differences are sketched here in broad brush strokes and extreme brevity (because of space limitations). I am aware that such comparisons harbor hidden Eurocentric assumptions about non-European knowledge systems (an issue discussed below in the context of cross-cultural learning; Ogawa, 2004a) but nevertheless, dichotomous comparisons can serve as a first step in a journey toward wisdom.

Indigenous sciences are guided by the fact that the physical universe is *mysterious* but can be survived if one uses *rational empirical means*. Western science is guided by the fact that the physical universe is *knowable* through *rational empirical means*. Both types of knowledge systems are guided by ontological facts, and both rely on empirical evidence gathered by experimentation and field studies. Indigenous scientific experimentation can take place over many generations, depending on the phenomenon, and it relies on natural changes rather than on laboratory manipulations.

Both indigenous and Western sciences employ rational ways of knowing, but their culture-laden rationalities differ to varying degrees in several ways (respectively):

- *social goals*: knowledge (ways of living) for survival and harmony with nature versus knowledge for its own sake, for economic gain, and for power over nature.
- *intellectual goals*: coexistence with the mystery of nature by celebrating mystery versus eradication of mystery by describing and explaining nature in ways familiar to Western scientists.
- *association with human action*: intimately and subjectively related versus formally and objectively decontextualized from normative prescriptions of human action.
- *notion of time*: circular versus rectilinear.
- *validity*: content validity as evidenced by tens of thousands of years of survival based on that content versus predictive validity that is the cornerstone of Western science.
- *general perspectives*: holistic, accommodating, intuitive, spiritual *wisdom* versus reductionistic, manipulative, mechanistic *explanations*.

The two knowledge systems share, however, a label they invoke to describe the other: *superstitious*. Scientists often describe indigenous science as superstitious because its spiritual dimension runs counter to their ontology of idealized rationalism (e.g., the world is made up of objective mathematical relationships). On the other hand, Lakota Elder Deloria (1992) stated, "What could be more superstitious than to believe that the world in which we live and where we have our most intimate personal experiences is not really trustworthy, and that another mathematical world exists that represents a true reality?" (p. 40).

Alaskan science educators spoke for many when they wrote, "Our experience with the Yupiaq culture in southwestern Alaska leads us to believe that such indigenous groups practice [indigenous] science in ways that [have] similarities to—and important and useful differences from—Western science, and that the worldview underpinning this indigenous vision of science has valuable implications for science instruction" (Kawagley et al., 1998, p. 133). Indigenous sciences hold valuable implications for the development of Western science itself (Cajete, 2000b). Yet there continues to be a power imbalance between the two types of knowledge systems in the context of school science.

STUDENT HETEROGENEITY

Not only is cross-cultural science education founded on respect and inclusion of students' indigenous science, but also its humanistic ideology implies responsiveness to student heterogeneity. Researchers have derived several ways to describe student differences. (Although these descriptions use categories, their intention is not to essentialize students.) The motivation for this research was to explain the underachievement and underrepresentation of certain groups of students in science courses with the hope of designing programs to achieve equity and social justice. The field of culture studies in science education avoids blaming students for their underachievement (within reasonable limits of students' human nature). Therefore, rather than label students as "at risk," we recognize that the teacher, the curriculum, the school, the community, and so on may be the at-risk elements in a student's underachievement. Hence we talk about a "vulnerable" student. A deficit model of learning is rejected in favor of an inclusive model that celebrates the non-Eurocentric cultural capital students bring to their science classroom (MacIvor, 1995; McKinley, 2005; Tobin, 1997).

Two interrelated research areas are considered here: first, the cultural similarities and differences between students' life-worlds and their science classrooms; and second, students' self-identities versus the school science identity expected of them by their teachers. Other descriptive categories arise in the context of cross-cultural teaching and learning (below).

Cultural Similarities and Differences

The degree of similarity or difference between a student's life-world (e.g., the culture of family and peers) and the culture of school was investigated by anthropologists Phelan et al. (1991). Their data suggested four categories that captured the full range of cultural similarities and differences high school students experienced between these two worlds. The categories are congruent, different, diverse, and highly discordant. In addition, Phelan et al. analyzed the ease with which students negotiated the transition from their life-world to the world of school, in terms of buying into the school's culture. Congruent worlds supported *smooth* transitions; different worlds required transitions to be *managed*; diverse worlds led to *hazardous* transitions (hazardous in terms of causing anxiety or psychological stress); and highly discordant worlds caused students to resist the transition, which therefore made the transition virtually *impossible*.

Inspired by this grounded theory, Costa (1995) embarked upon an ethnographic study into the ease of student transitions from their worlds of family and friends to the worlds of school and school science. She related students' consistent success in high school chemistry with how easily they appeared to negotiate transitions into those classes. A five-category scheme of transitions into school science emerged from her data, to which I added a sixth category based on my own research (the second category in the list below; Aikenhead, 2001b). The names of the categories are logically explained by Costa (1995). The categories are not stable but depend on context. I also expanded Costa's focus on student achievement by considering the relevance of school science to students, their self-esteem, and their image of themselves as science students (Aikenhead, 2001b). The transitions experienced by students range from effortless (invisible) to highly problematic (painful). The scheme is summarized here.

- *Potential Scientists*: whose transitions are smooth because the cultures of family and friends are congruent with the cultures of both school and Western science. This leads to a meaningful understanding of science content because students' worldviews, self-images, and lifestyles resonate with the world of Western science. Transitions are invisible; they do not exist for these students.
- *"I Want to Know" Students*: whose transitions are hazardous but well worth the adventurous risk because their worldviews, self-images, and lifestyles resonate with the world of Western science even though scholarly barriers of some sort prevent them from becoming Potential Scientists (e.g., math ability). Hazards exist, but students want to know. This leads to a modest yet effective understanding of science content.
- *Other Smart Kids*: whose transitions are manageable because the culture of family and friends is congruent with the culture of school but inconsistent with the culture of Western science owing to these students' lack of personal interest in science. These students do not fit the worldview, self-image, and lifestyle they associate with Western science, but they do have strong identities related to school

success, so they achieve high marks in science either by understanding the content or, more often, by playing Fatima's rules.

- *"I Don't Know" Students*: whose transitions tend to be hazardous because the cultures of family and friends are inconsistent with the cultures of both school and Western science. Students do not want to "look stupid" in the eyes of peers or teachers (i.e., a hazard), so they persevere into a superficial understanding of science content, usually by playing Fatima's rules.

- *Outsiders*: whose transitions are virtually impossible because the cultures of family and friends are discordant with the cultures of both school and Western science. Their worldviews, self-images, and lifestyles are at odds with their understanding of Western science. They usually drop out physically or intellectually to protect their self-image and avoid stress and anxiety.

- *Inside Outsiders*: whose transitions are frustratingly difficult or impossible because the cultures of family and friends are irreconcilable with the culture of school but are potentially compatible with the culture of Western science. This happens because of institutional discrimination against the student in spite of the student's personal interest in understanding science content.

This scheme is illustrated in other studies (Aikenhead, 1996, 2001b; Aikenhead & Jegede, 1999; Cobern & Aikenhead, 1998).

By describing similarities and differences in general terms, the scheme identifies a variety of culture clashes that occur in a science classroom. It also helps to explain why so few students respond favorably to traditional science content (Chapter 3). Multicultural researchers Brown et al. (2004) concluded, "Learning to participate in the cultural practices of school science is a complicated endeavor for many students to accomplish" (p. 7). Only Potential Scientists and "I Want to Know" Students (a very small minority) would find wish-they-knew science relevant to their life-world, that is, relevant to their cultural self-identities. Cultural relevance for other students, the large majority, would look very different (Chapter 3). Their self-identities hold a key to making school science more educationally sound.

Student Self-Identities

In a science classroom, culture transmission leads to culture acquisition (Wolcott, 1991), a process that engages students' self-identities and transforms aspects of who they are and what they want to become (Brickhouse et al., 2000; Hammond, 2001; Kozoll & Osborne, 2004; Nicol et al., 2004). Students' self-identities are not meaningfully engaged, of course, by playing Fatima's rules.

The scholarly literature describes student self-identity in multifarious ways (Brown et al., 2004), for example, in terms of one's natural characteristics (e.g., color markers), one's formal institutional allegiances, one's discursive habits, and one's affinity to others who share common interests and experiences (Gee, 2002). Thus a student's self-identity is composed of many identities; some are personal because they are coherent across contexts owing to a student's historical past and anticipated future, whereas others are group- or culture-related because they are based on race, class, gender, ethnicity, language, religion, and so on (Gagné, 2004). Some self-identities are changeable through social interaction. Different identities develop when we take on different "ways of talking, acting, being in the world, describing oneself, or relating to others" (Carlone, 2004, p. 396). The potential to describe student heterogeneity is limitless when we consider the many identities one student possesses, summarily referred to as his or her self-identity.

Meaningful learning in school science has been conceived in terms of students' forming new self-identities (Brickhouse, 2003; Carlone, 2004; Thompson, Smith, & Windschitl, 2004). "We need to consider how learning science can change students' identities by changing their ability to participate in the world" (Brickhouse, 2001, p. 288). In other words, to learn science meaningfully is to engage in identity work. Some science education researchers try to foster positive school science identities so students will learn to talk and think like scientists (Brown, 2004; Kozoll & Osborne, 2004), referred to earlier as enculturation or assimilation into scientific disciplines. If students begin to talk like scientists, then others will identify them as competent science students, which they may or may not be according to their other self-identities.

But this attempt at enculturation or assimilation into scientific disciplines often fails when students have self-identities as, for instance, an artist, an Aboriginal person, or a caring religious person (Brickhouse & Potter, 2001; Brown, 2004; Calabrese Barton, 2002; Chinn, 1999; Gagné, 2004; Seiler et al., 2001). Clashes between students' self-identities and school science identities provide rich detail for understanding Costa's (1995) categories described earlier. For instance, students may become Outsiders in order to resist instruction, thereby preventing their self-identities from being changed by the dominant culture of Western science (Lee, 1997; McKinley, 2003; Norman et al., 2001). But underachieving students (also Outsiders, perhaps) may view school science as irrelevant because it does not help them express themselves, build on or develop new identities, develop or maintain relationships with others, or learn about themselves and others (Thompson et al., 2004).

Some students endeavor to develop self-identities as good students, that is, achieving high marks (Carlone, 2004; Nicol et al., 2004). For them, being a good student will take preference over developing a school science identity, which may cause these students (Other Smart Kids) to avoid difficult science classes for fear of getting low marks. On the other hand, Costa (1997a) discovered that an identity as a high-status student (defined by the culture of a school) caused some students (Other Smart Kids) to enroll in a high-status honors chemistry course, where they played Fatima's rules to succeed within the school culture but avoided forming a chemistry identity expected by the chemistry teacher.

A school science identity or a good student identity may prove disastrous to students whose peers find these identities socially unacceptable ("It's not cool."), causing these students to be unwilling to engage even in science discourse (Brown, 2004). Brown's ethnographic data suggested a pattern in the degree to which students subscribed to a "scientific discursive identity" (a cultural identity related to one's participation in scientific discourse and other scientific practices). He proposed four categories of students ranging from "proficiency status" to "opposition status."

When researchers in culture studies reported positive results in their attempts to persuade students to take on a school science identity (often delimited to a discursive scientific identity), a closer examination of their project reveals that the content and instruction did not conform to traditional school science but instead exemplified humanistic school science (Brickhouse, 2003; Calabrese Barton, 2004; Gagné, 2003; Kozoll & Osborne, 2004; Rahm, 2002; Seiler et al., 2001). To achieve the researchers' objectives a humanistic approach was required.

Research into students' self-identities dramatically clarifies the educational inadequacies of traditional school science (Chapter 3). However, a humanistic approach to school science does not solve students' identity clashes with a school science identity. Instead, a humanistic approach respects and values the diversity of student identities formed in part from local knowledge and social interaction patterns in the community. A cross-cultural (humanistic) perspective nurtures students' self-identities as savvy citizens capable of critically interacting with science-related events and issues in their everyday world.

Rather than insist that all students develop discursive scientific identities (Brown, 2004; Carlone, 2004), educators could help some students develop discursive *expertise* in Western science (being able to communicate in the culture of Western science without forming scientific self-identities). This could occur via "anthropological instruction" from which students learn science in the same manner an anthropologist learns a foreign culture (Aikenhead, 1997). At the same time educators could help students enhance their self-identities in more autonomous and realistic ways (e.g., becoming a better commercial gardener or a more competent participant in debates over genetically modified organisms). This distinction between identity and expertise reflects the difference between the enculturation of students into scientific disciplines and the enculturation of students into their local, national, and global communities, respectively, described in Chapter 3 as cultural relevance.

CROSS-CULTURAL TEACHING AND LEARNING

A very limited amount of research has targeted cross-cultural teaching and learning because most science education projects have tended to be missions for improving marginalized groups' participation and achievement in science education rather than sites for publishable research.

Cultural Border Crossing

Cultural border crossing is a metaphor that captures the act of negotiating the transition from, for instance, a student's home culture to the culture of school science. The idea can be located in the work of a number of researchers, but it was Giroux's (1992) *Border Crossings: Cultural Workers and the Politics of Education* that promised to connect educational anthropological research with teacher practical knowledge (Chapter 5) and therefore with classroom practice (Aikenhead, 1996; Aikenhead & Jegede, 1999; Jegede & Aikenhead, 1999; Pomeroy, 1994). Giroux contrasted modernist and postmodernist views on education. Modernism, he contended, defines borders and locates people within those borders along with the static social and political power afforded each location. On the other hand, postmodernism, like cross-cultural science education, acknowledges that people have multiple self-identities and encourages them to live in a world of border crossings between many locations (Lugones, 1987). Border crossing is a politically neutral phrase that acknowledges two different cultures. Erickson (2004), however, interpreted the phrase as being politically charged and posited "boundary crossing" as his neutral phrase. In this book, border crossing signifies Erickson's (2004) boundary crossing. It assumes a capacity to think differently in various cultures (e.g., in the culture of Western science), a capacity to talk about those differences (metacognition not normally prevalent in elementary students, thereby making border crossing discussions inappropriate in elementary school science), a capacity to feel at ease in a less familiar cultural context, and a capacity to resolve conflicting beliefs.

The ease of border crossing in and out of school science can be described in many ways, for example, Costa's (1995) categories (presented earlier). She associated the ease of border crossing with students' success in school science. Krogh and Thomsen (2005) associated border crossing with students' attitude toward physics and their choice to study more physics courses. Working within the quantitative paradigm, Krogh and Thomsen's multiple regression models posited three powerful predictors for ease of border crossing: students' physics self-identity, physics' reputation for not being strange or boring, and teachers' expressed interest in students as persons.

Culture-based conflict that may arise in school science can motivate some students to develop psychological strategies to ameliorate the stress associated with conceptual cultural conflict, a process called "collateral learning" (Aikenhead & Jegede, 1999; Ezeife, 2003; Jegede, 1995; Tsai, 2001b). At one extreme of collateral learning, students compartmentalize two cultural knowledge systems (e.g., indigenous and Western sciences), so there is no interference between the two, whereas at the other extreme, students resolve the original conflict in one of several possible ways described in Aikenhead and Jegede (1999).

The notion of cultural border crossing has guided the development of teaching materials in some Canadian Cree and Dëne communities and in some Aotearoa New Zealand Mäori schools (Chapter 4). Sutherland (2005) investigated the ease of border crossing by Cree students and found that their ease was associated with their intrinsic motivation and with their metacognition of two knowledge systems (indigenous science and Western science), that is, a type of collateral learning. The intrinsic motivation arose in part from the relevance of the contextualized science content taught at their school, from the culturally sensitive instruction provided by their Aboriginal science teacher, but also from a larger constellation of factors known as student *resiliency*: the ability to thrive academically, form new self-identities, and improve competencies in the face of adverse circumstances or obstacles (Sutherland, 2005). Resiliency largely originates within students themselves and with their positive relationships with friends, family, and community members past and present (Lyons, 2003; McKinley et al., 2004).

Science teachers can help students negotiate difficult border crossings, but this help is normally absent in science classrooms because teachers are unaware that cultural borders exist for their students (Contreras & Lee, 1990; McKinley et al., 2004; Phelan et al., 1991). In fact, most science teachers are unaware of the cultural nature of the school science they teach

(Aikenhead & Huntley, 1999; Aikenhead & Otsuji, 2000; Haidar, 2002; McKinley, 2005). Their students must rely on resiliency to negotiate cultural border crossings into school science on their own. Students in South Africa, however, had little problem crossing cultural borders into and out of school science when the culture of the classroom and the politics of the curriculum encouraged border crossing (Malcolm & Alant, 2004).

Aboriginal educators conferred the role of "culture broker" on teachers who attempted to ease student border crossings into and out of school science (Aikenhead, 1997; Michie, 2004; Stairs, 1993/94). Culture brokers act in several ways: They acknowledge that a border exists, they motivate students to cross it, they employ language of both the students' culture and the culture of Western science, they explicitly keep track of which culture comprises the context at the moment, and they help students resolve cultural conflicts that may arise (Aikenhead, 1997; Chang & Rosiek, 2003; George, 1999b; Jegede & Aikenhead, 1999; Rollnick & Rutherford, 1996; Sutherland & Dennick, 2002). Culture brokers motivate students, for example, by developing a relationship with them, by understanding the specific history of the students' culture, and by holding high expectations for them (McKinley et al., 2004). The role of culture broker is very similar to multicultural pedagogy (Erickson, 2004; Hines, 2003); that is, a culture broker reframes the acquisition of relevant Western science as an appropriation of Western culture for utilitarian purposes rather than as the correct way of knowing about the world (Chapter 1). Culture brokers will make the ontology of the Western colonizer explicit in their classrooms (e.g., clarifying the mathematical idealization of the physical world), thereby providing students more freedom to appropriate parts of Western science without embracing Western ways of valuing nature (Cajete, 1999; Hodson, 2001; Krugly-Smolska, 1994; MacIvor, 1995), an appropriation called "autonomous acculturation" (Aikenhead, 1997).

However, student participation and achievement in school science is undermined by low student self-expectations; low teacher efficacy and expectations; inadequate subject matter knowledge, pedagogical flexibility, and cultural knowledge held by teachers; stereotyping students; and rigid science curricula and school timetables (Brand, Glasson, & Green, 2004; McKinley, 2005).

One research study investigated the role of a culture broker teacher in the United States. He detected conflicts between his biology course content and some Hmong immigrant students' cultural beliefs (Chang & Rosiek, 2003). The case study revealed tensions, uncertainties, and the teacher practical knowledge that led to his anticolonialist reaction to their conflict. Further research needs to clarify various roles of a culture broker in science education (Michie, 2004).

Yet research into cross-cultural science teacher education has established some key findings related to culture brokering (ANKN, 2003; Atwater & Crockett, 2003; Bishop & Glynn, 1999; Chinn, 2004; Luft, 1999; Rowland & Adkins, 2003). Teachers need to articulate and reflect on their own personal culture before they can learn about their students' cultures, and they need to immerse themselves in their students' cultures cognitively, metacognitively, and emotionally before they can develop their own unique classroom culture to support their role as an effective culture broker. This classroom uniqueness arises from a combination of social interaction patterns, language usage, preferred ways of teaching and learning, and beliefs about success that determines motivation (Tharp, 1989). Power relationships within a classroom alter accordingly (Aikenhead, 2001a; Gagné, 2004). For Aboriginal students, of course, the best culture broker would be an Aboriginal science teacher (McKinley, 2001; Sutherland, 2005).

Chapter 5 identified interactivity as a critical feature to successful instruction for most students. To achieve interactivity requires teacher sensitivity to various ways of learning and to cultural interactions in the community. There is no such thing as an Aboriginal learning style, but Aboriginal students tend to express culturally related, recurrent strengths in their ways of learning (Hughes et al., 2004). These context-dependent strengths include "global more than analytic; imaginal more than verbal; concrete more than abstract; reflective more than trial-and-feedback, in school; and contextual more than decontextual" (p. 183). Effective culture brokers draw upon students' strengths to help them expand their repertoire of ways of learning. Other teaching methods that appear to support Aboriginal strengths include modeling an adult, guided practice, and instruction "in the land" (Aikenhead, 2000a; Kawagley et al., 1998; Rowland & Adkins, 2003; Stephens, 2003). These methods appeal to various non-Aboriginal students as well.

In a case study of a student teacher placed in a cultural setting very different from her own White middle-class culture, several cultural borders were identified (Luft, 1999). The student teacher negotiated some hazardous borders successfully (e.g., into her students' culture and into the school's culture), but not others (e.g., into the science department's culture that emphasized a scientist-oriented transmission of content). Luft's research, along with that of others (Atwater & Crockett, 2003; Ezeife, 2003; Mulholland & Wallace, 2003), demonstrates the reflexivity available to student teachers and their supervisors when they use border crossing as one lens among many to examine classroom practice and enhance professional development.

Integration of Indigenous and Western Sciences

A curriculum that integrates a student's indigenous science with relevant Western science should attain at least three goals: to help students cross a cultural border into and out of school science, to nurture and expand students' self-identities, and to prepare students to live proficiently in both their indigenous and Western worlds (Bajracharya & Brouwer, 1997; Cajete, 1999; George, 1999a, 1999b; Kawagley et al., 1998; MacIvor, 1995; McKinley et al., 1992; Rowland & Adkins, 2003; Snively & Corsiglia, 2001).

Whether the two knowledge systems are similar or different depends on the specific content. George's (1999a) research in a village of Trinidad and Tobago ascribed four categories to a continuum from one extreme, where each knowledge system could explain the content of the other system, to the opposite extreme, where their differences caused the type of cognitive conflict that requires collateral learning. Integration of indigenous and Western sciences will vary depending on the content to be taught.

A number of years ago, a popular Western science ecology curriculum, Project Wild, began to be transformed into a Canadian Aboriginal curriculum in order to integrate both sciences. After several days of frustration, the Aboriginal participants withdrew. On their own, however, they produced *Practising the Law of Circular Interaction* (Saskatchewan Indian Cultural Centre, 1993), an indigenous ecology curriculum. The differences in epistemology, ontology, and axiology between both knowledge systems were so discordant they necessitated a fresh independent start. A simple transformation from one system to another is often not feasible (McKinley, 2005). Thus integration can fail depending on the context.

Successful integration has been achieved, for instance, by the Alaska Native Knowledge Network's *Spiral Pathway for Integrating Rural Alaska Learning* (ANKN, 2004b), by Read's (2002) *The Kormilda Science Project* in Australia, by some Canadian Cree and Dëne schools' *Rekindling Traditions* (Aikenhead, 2000a), and by a bilingual network of Mäori schools in Aotearoa New Zealand (McKinley, 2005). In Alaska, Native American students' standardized science test scores uniformly improved over 4 years to meet with national averages in classrooms where teachers achieved integration by teaching ANKN science modules (Barnhardt et al., 2000). In her analysis of the *Rekindling Traditions* materials, Gagné (2004) found that its lessons introduced to students "issues of identity, knowledge and power" (p. 19), and that indigenous and Western sciences were presented as local systems of knowledge to be used in particular contexts without necessarily assimilating the values each system conveyed.

In the Yupiaq and Inuit nations of North America's far north, integration is specified at the policy level (ANKN, 1998, 2004a; Inuit Subject Advisory Committee, 1996; Kawagley, 1995; Stephens, 2000). A case study into the integration of the Canadian Nunavut document *Inuuqatigiit: The Curriculum from the Inuit Perspective* (Inuit Subject Advisory Committee, 1996) into a grade 7 science classroom identified advantages and challenges when the integration was the sole responsibility of the teacher (DeMerchant, 2002). The advantages included relevant contexts for student learning, a base from which a teacher can relate to students, respect for students as holders of local knowledge, and an increase in student self-esteem. Some challenges were time-tabling problems, greater preparation by teachers, and incompatibility with external exams.

The idea of integrating indigenous and Western sciences in a classroom was expanded into a multiscience perspective with the addition of a student's personal science into a cross-cultural classroom (Hammond, 2001; Ogawa, 1995). Our understanding of integration expanded even further when Ogawa (2004a) described integration from a Japanese insider's point of view, different from a Eurocentric outsider's point of view. He argued persuasively that indigenous science is more than a single knowledge system comparable to Western science. Instead it is "a body of stratified and amalgamated knowledge and cosmology" (p. 1) that has evolved or was borrowed from earlier cultures. Ogawa's argument illustrates a translation issue mentioned above, in which "knowledge" in an Indo-European language is interpreted as "ways of living" in many non–Indo-European languages, such as Japanese in this context (Kawasaki, 2002). Ogawa's argument also illustrates how Western educators are often unable to escape their Eurocentric Cartesian habits of mind, and therefore we unconsciously attribute our own ontology and epistemology to people indigenous to different cultures (e.g., Japanese people). Ogawa's research described ways to analyze the complexity of Japanese contemporary indigenous science (not to be confused with the Aboriginal science of the Ainu who reside in Japan). Particularly fruitful was Ogawa's natural link between ways of living and a person's identity formation (that person's cultural self-identity). Ogawa's work elucidates richness in indigenous student learning in cross-cultural science classrooms.

We should not assume that students' learning conforms to the integrative structure of their teacher's cross-cultural teaching. Students will integrate indigenous and Western sciences in unique ways that make sense to their worldviews, self-identities, personal science, and indigenous science. Some people accomplish this holistically; they do not identify their world as a multicultured world, thus they do not identify with a bicultural description of it (Brandt, 2004). For them, border crossing does not adequately describe their experiences in school science. But others are quite conscious of their role in a multicultured world (Brown, 2004; Jegede, 1995; Lugones, 1987). Cross-cultural science teaching can only make indigenous and Western sciences *accessible* to students, cognitively, emotionally, and culturally. How students individually integrate the two, if at all, is always their prerogative. Further research on this phenomenon is required.

Researchers have noted some student ambivalence or frustration over "two-way" (bicultural) learning that includes Western science (Baker & Taylor, 1995; Brown, 2004; George, 1999a), but researchers have also documented its transformational effects (McKinley, 2003). In an American Pueblo setting, Peshkin (2000) blamed student ambivalence and frustration not on schooling but on a lack of appropriate role models in the Pueblo community, models that would demonstrate two-world integration. Peshkin, however, did not consider the effects of the dominant world's discrimination and racism on students who dare succeed at school science, effects found in other studies (Bianchini et al., 2000; Chinn, 1999; McKinley, 2003; Sutherland & Dennick, 2002). The Pueblo students may have been avoiding future stress and anxiety in their lives. Because gender discrimination within the scientific enterprise was an effective researchable topic taken up by feminist scholars (Bianchini & Solomon, 2003; Scantlebury, 1998), racism may be a potentially fruitful target of research to explain some of the underachievement and underrepresentation by Aboriginal students in the academy and science-related professions.

The Role of Language

McKinley et al. (1992) enriched the idea of integration when they argued for students to learn both indigenous and Western sciences in a student's Aboriginal language. This led to research into translating the Aotearoa New Zealand science curriculum into the Māori language (McKinley, 1996) and to research into the effect of non–Indo-European languages on student learning (Kawasaki, 1996, 2002; Lee & Fradd, 1998; Lynch, 1996; Ogawa, 1997; Sutherland & Dennick, 2002).

Today's research literature amply illustrates the major problem of students' everyday language interfering with their conceptual learning in school science; for example, "force" in their everyday culture usually means "momentum" in the culture of physics (Aikenhead, 2000b). This cultural divide widens when students' language is not the language of instruction (Contreras & Lee, 1990; Lee, 1997; Rollnick & Rutherford, 1996) and is much wider when students speak non–Indo-European languages, such as Māori (McKinley et al., 1992), Mi'kmaq (Berkowitz, 2001), or Japanese

(Kawasaki, 1996). Incommensurability creates hazardous border crossings. Imagine the adversity trying to learn Western scientific knowledge when *knowledge* intuitively means "a way of living" to a student. Imagine the challenge learning abstract scientific concepts when your Japanese language does not possess such abstract nouns (Kawasaki, 2002). Language is more than terminology; it is a cultural repository of worldviews that teaches people what to assume, how to think, and how to socially interact (Lemke, 2001; McKinley, 2005). Thus many non-Western students risk assimilation into Western culture (i.e., colonization) if they become enculturated into Western science by forming, for example, a scientific discursive self-identity. Resisting such enculturation by refusing to engage in scientific discourse can be, in certain circumstances, a rational act of self-preservation and cultural survival.

When an Aboriginal nation loses its language, it loses a large part of its culture (Kawagley et al., 1998; McKinley, 2005). Consequently, language preservation is a high priority in indigenous education. Senior secondary students in Mäori immersion science classes, however, expressed a preference for English science instruction because people use English to communicate in science and technology outside of their school (McKinley et al., 2004). Further research is needed to clarify an effective match among, for example, language of science instruction, grade level, family culture, student identities, and types of students described by Costa's (1995) scheme.

Much of the science education research on second-language instruction has been conducted in elementary grades (e.g., O. Lee, 2002), a venue beyond the scope of this book. But Lee and Fradd's (1998) "instructional congruence" model can be applied to cross-cultural teaching and learning for older students as well.

Even when Western science is taught to Western students in their mother tongue, students often find the genre of scientific writing a hazardous cultural border to cross (Knain, 2005; Lemke, 2001).

CONCLUSION

The research synthesized in this chapter points to pockets of positive results in this emerging complex field of culture studies in science education. Student interest, self-esteem, achievement, and empowerment can be augmented by a cultural approach to school science. This holds promise for equity and social justice goals. However, most science teachers do not appreciate the fact that the vast majority of their students, those who feel school science is a foreign type of culture, must cross cultural borders and form new self-identities to achieve the goals of the traditional science curriculum (or they must play Fatima's rules). If teachers themselves formed new professional identities to embrace teaching science as cultural transmission sensitive to the dangers of cultural assimilation (colonization), then the educational soundness of their work would be more defensible. But as we saw in Chapter 5, forming new professional identities can be very difficult.

A considerable array of issues in culture studies remains to be explored analytically and empirically. Most critical are issues regarding cross-cultural (bicultural) learning: first, for students who need an enculturation into their local, national, and global communities; and second, for students who want an enculturation into Western science disciplines without losing their cultural self-identities.

Curriculum policy in support of humanistic cross-cultural learning is already in place for the most part, but teaching materials are meager. Teacher development requires considerably more attention, but we can be encouraged by recent action research studies (e.g., Barker, 2004; Chinn, 2004; Sutherland & Tays, 2004). Research into students' cross-cultural learning has begun, but many questions remain.

Perhaps if greater attention were given to making border crossings into and out of school science more manageable, researchers could discover in greater depth students' willingness to develop new self-identities relevant to the well-being of their community's culture. Student resilience and teacher culture brokering are fruitful avenues of research to pursue, as is Brandt's (2004) research program into how non-Western science-proficient students integrate personal, indigenous, and Western sciences when they form self-identities within the culture of Western science.

"Native people have the double advantage of knowing and living in two cultures" (Galindo & Barta, 2001, p. 54). But this advantage does bring with it unique challenges to students and to their science teachers who strive to become cultural workers (Giroux, 1992).

8

Moving Forward

In this chapter I do not attempt to summarize the many conclusions and recommendations posited throughout the book. Instead I discuss past and future research in terms of advancing humanistic perspectives in school science, in the context of educationally and politically driven agendas that motivate research programs. The underlying theme is: What counts as science education today?

CREDIBILITY

The post–World War II renaissance in humanistic school science understandably forced researchers to establish its credibility. Science teachers, science educators, and policy makers had to be convinced. The research reviewed and synthesized throughout this book provides strong and substantial evidence in support of the educational soundness of humanistic school science. It is undoubtedly credible on educational grounds. Therefore, we can stop doing research that simply shows a humanistic approach can work successfully under reasonably favorable conditions (e.g., Schwarz-Bloom & Halpin, 2003).

This evidence-based success has influenced some national curriculum policies to take on a humanistic perspective (Aikenhead, 2000c; Ratcliffe et al., 2003), yet it has had little impact on classroom practice (Bybee, 1993; Cheek, 2000; Gallagher, 1998; Gaskell, 2003). Ratcliffe et al. (2003) in their work with teachers concluded, "Unless research evidence, including that from highly regarded studies, is seen to accord with experience and professional judgement it is unlikely to be acted on" (p. 21). How teachers enact a humanistic approach is not so much an educational issue as it is a political issue. Research programs need to address the political dimensions of humanistic school science to improve its political credibility (Barrett & Pedretti, 2004; Fensham, 2002).

To gain political credibility humanistic researchers must target political change (Aikenhead, 2000a; Calabrese Barton, 2002; Elmore, 2003; Medvitz, 1996; Solomon, 2002; Tobin & McRobbie, 1996; Venville et al., 2002). The most effective research will explore *the interaction of political power with research, policy, and practice* (Alsop, 2003). Only a few studies have investigated such an interaction (Boydston, 1999; Carlone, 2003; Fensham, 1998; Hart, 2001; Östman, 1996; Rigano & Ritchie, 2003). "When power relations change in society and in the community of education, then what counts as science education would or should shift" (Östman, quoted in Fensham, 2004b, p. 110). We need to learn more about political power at the policy level, at the teacher education level, at the school level, and at the student level where politically savvy actions by students have made a difference (Carlone, 2003; Eijkelhof & Kapteijn, 2000; McRobbie & Tobin, 1995; Nicol et al., 2004; Wallace & Louden, 1998).

A humanistic perspective will significantly alter the culture of school science and the culture of schools (Chapter 5). This entails renegotiating values and concepts related to the status of school subjects. For example, high school preprofessional training courses for science-proficient students (i.e., college preparatory science courses) might be re-conceptualized administratively in line with other preprofessional training courses, such as business education or technical vocation classes.

We can anticipate a visceral negative reaction from pipeline enthusiasts as they position themselves to protect the status quo curriculum and their privileged status. They will likely misconstrue our political initiatives as an assault on Western science itself (Blades, 1997; Cross, 1997b). This negative reaction is not an educational problem but a political one, which itself is ripe for research by science educators.

Because teachers are pivotal to the politics of the taught curriculum, continued research into understanding their successes and failures at implementing a humanistic perspective will be valuable (e.g., Bartholomew et al., 2004; Tal et al., 2001). The complexities of teaching are no longer expressed as relationships among variables captured by statistical analysis but are now appreciated using heuristic models such as teacher practical knowledge, educational worldview, teacher context knowledge, and culture brokering (Chapters 5 and 7). Fine-tuning these types of schemes affords deep insight into what is actually involved in changing a taught curriculum.

Because student teaching or internship in schools is critical to novice teachers' loyalty to a humanistic perspective, research in this vulnerable area will be particularly fruitful (e.g., Pedretti et al., 2004). The results must increase the proportion of humanistic science teachers in the profession rather than allow the current system to undermine humanistic preservice teacher education programs (Chapter 5).

Caution is advised, however. The political credibility of humanistic school science can be jeopardized if humanistic policies and actions are too ambitious or radical to succeed in the prevailing political climate of a particular educational jurisdiction (Black, 1995). How can we influence that political climate? This crucial research question must be incorporated into future research programs if humanistic approaches to school science are to advance (e.g., Aikenhead, 2002b).

The political credibility of a humanistic approach will be enhanced by multidimensional research (described later) that co-opts diverse stakeholders of social privilege, over a long period of time (Anderson & Helms, 2001). Accordingly, research projects should involve clusters of science teachers conducting the research, and projects should involve people who are empathetic to a humanistic perspective and who are stakeholders holding greatest power over curriculum policy (e.g., university science professors, executives of major local industries, and other leaders).

FUTURE RESEARCH STUDIES

To investigate the interaction of political power with research, policy, and practice, for the purpose of facilitating changes to school culture and classroom practice, researchers must revisit the aims of science education currently encased in a 19th-century ideology (Chapter 2). One fundamental dilemma must be addressed by each research project: Does it aim to enculturate students into their local, national, and global communities (as some school subjects do), or does it aim to enculturate students into scientific disciplines (Chapters 3 and 7)? The choice is essentially between cultural relevance and playing Fatima's rules.

Many science educators believe that enculturation into scientific disciplines is necessary to supply the pipeline with a sufficient number of students. These educators should not ignore the research that places responsibility for supply on university undergraduate programs, where the exodus from the pipeline is twice that in high schools in the United States (Chapter 3). We must recognize that when educators ignore such evidence-based findings, they are participating in a political act based on their personal values and ideologies. Similarly, I participated in a political act in this book by advancing the values and ideologies endemic to a humanistic perspective.

One of the most promising but most challenging directions for future research is what I would call "consensus-making R&D." This is action research on the grand scale of deliberative inquiry accompanied by curriculum implementation, within a large educational jurisdiction and drawing upon a broad array of stakeholders judiciously chosen so the political elite is represented but its status quo science education is actually discussed and renegotiated. During deliberative inquiry meetings (Chapter 3), research concerning major failures of the traditional curriculum will be scrutinized, research concerning successes at learning science in nonschool settings will be debated, and research concerning relevance will help clarify or realign participants' values and loyalties. The aim is to transfer to school science the successes of people who learn Western science outside of school (lifelong learning). This consensus-making R&D needs to take place over several years so that further research is conducted at the request of the deliberative inquiry group. Significant change to school science and teacher education programs will be measured by decades of ever increasing implementation, as vicious cycles favoring traditional school science are eroded by an evolving core of humanistic science teachers.

Consensus-making R&D will have greater impact on classroom practice the more culturally transformational it is (McKinley et al., 2004), the larger its scale is (Coburn, 2003; Elmore, 2003; SCC, 1984), and the more it embraces a combination of research paradigms (discussed later). Worthwhile research could investigate, for example, the influence of various participant–stakeholders in the consensus-making R&D. How were they chosen? What were their anticipated versus enacted roles? What actor-networks did they bring to the deliberations and what networks developed as a result of the deliberation (Carlone, 2003; Gaskell & Hepburn, 1998)? R&D on roles and actor-networks could be a primary focus of a deliberative inquiry project, forging new roles and networks to enhance a clearer and more politically endorsed humanistic perspective.

Preliminary small-scale studies can also make a significant contribution to future research: "Rather than viewing the powerful sociohistorical legacy of science as an oppressive structure that limits the potential of reform, we can view the meanings of science in local settings as partially fluid entities, sometimes reproducing and sometimes contesting sociohistorical legacies" (Carlone, 2003, p. 326). But small-scale studies will lose significance unless they explicitly embed themselves in a larger, articulated, politico-educational agenda for humanistic school science.

"American educators tend to assume that [Western] science is a natural part of at least Euro-American students' culture" (Cobern, 1996b, p. 602). This assumption is clearly false, as Cobern's research demonstrated. Traditional school science is educationally irrelevant for most students, and teachers' attempts to make it appear relevant are undermined by students' playing Fatima's rules (Chapters 3 and 6). Traditional science content makes sense only to students who possess scientificlike worldviews, in short, to students who want to work at forming self-identities related to Western science and technology, described in Chapter 7 as Potential Scientists and "I Want to Know" Students. For all the other students, traditional school science has not engaged them and it never will. For them school science is educationally bankrupt but continues to function on the currency of its political credibility. After investigating why so many clever students no longer took optional science courses in high schools today, Lyons (2003) suggested researchers ask, Why should they?

Research studies into Fatima's rules played by various types of students and science teachers (e.g., McRobbie & Tobin, 1995) could reveal interesting influences by, for instance, high-stakes testing, school policies, and student self-identities. Without this type of research placed on our agenda, Fatima's rules will remain the proverbial "elephant in the room" that inhibits advances toward humanistic school science.

Some humanistic researchers are exploring how students can develop a positive relationship with Western science as a basis for lifelong learning (Rennie & McClafferty, 1996). Other key educational issues to be explored include the following: How do students struggle and cope with worldview clashes in their science classrooms? What are successful classroom practices for Aboriginal students and how can those practices help most non-Aboriginal students? and How do students come to value and appreciate specific humanistic content? A pressing political research question is, Who will allow students to learn science from a humanistic perspective?

Canonical science content does have its place in humanistic school science, but that place is subordinate to many types of relevant science (i.e., have-need-to-know science, science-as-culture, etc.; Chapter 3) for the purpose of enculturating students into their local, national, and global communities. The "mental training" argument supporting traditional science content (Chapter 2) works equally well in support of have-need-to-know science and science-as-culture, all for the purpose of lifelong learning. For pipeline enthusiasts, however, this altered mental training argument will likely become a political issue (e.g., the school's status derived from student scholarships at prestigious universities), not an educational one. An appropriate response will require political action, not educational arguments based on research.

More research into the various types of relevant science will be helpful. How do science-proficient students actually use school science content in their everyday lives (if at all), compared with science-shy students coping with similar situations? Who in the community is engaged with science and technology, and how? What can students learn from them, directly or indirectly? Scaled-up versions of work by Law (2002) and by

Symington and Tytler (2004) are needed to delineate have-cause-to-know science or science-as-culture valued by politically positioned citizens.

The educational issue here is one of engaging students' self-identities (Chapter 7). How can school science do this, or is it even feasible? Answers to these types of questions will come most easily from community-based participatory research (e.g., Bouillion & Gomez, 2001; Cajas, 1998; Cajete, 1999; Calabrese Barton, 2004; Dori & Tal, 2000; Jiménez-Aleizandre & Pereiro-Muñoz, 2002; Kawagley et al., 1998; Roth & Lee, 2004). Unfortunately, this methodology is feasible in only a limited number of communities. On the other hand, reasonable classroom *simulations* of community, national, and global events have been successful to some extent (Chapter 6) and are welcomed by a far larger number of communities.

RESEARCH PARADIGMS AND METHODOLOGIES

It is convenient to reflect on humanistic science education research in terms of five research paradigms. Four refer to Eurocentric educational research (Calabrese Barton, 2002; DeBoer, 1991; Eisenhart, 2001; Ryan, 1988): historical, quantitative, interpretive (qualitative), and critical-theoretic (transformative/emancipatory). A fifth paradigm is currently emerging from a field of non-Eurocentric scholarship: an Aboriginal way of generating educational knowledge (Smith, 1999; Wilson, 2003). Each paradigm is summarized here to sketch the multidimensional character of politically credible research that will advance humanistic school science.

Although only a small number of science educators have conducted research in the historical tradition, their contributions to ongoing debates in science education have been valuable, evidenced by Chapter 2. Not only do historians contextualize issues for greater understanding, they challenge our thinking by articulating alternative perspectives.

Some science educators feel comfortable in the role of disinterested observer (the quantitative paradigm). Quantitative research draws upon positivistic notions about the measurement of outcomes evaluated against expert judgments or against criteria from academic theoretical frameworks. The quantitative paradigm dominated science education at a time when a good science curriculum was deemed to be a teacher-proof curriculum, and when inservice programs simply transmitted the new curriculum's philosophy to passive teachers (Solomon, 1999a). It found ultimate expression in the Harvard Project Physics research program (Welch, 1973). The paradigm's philosophy of conforming to preestablished frameworks and norms is being revitalized today by innovative research in Europe

(Krogh & Thomsen, 2005) and by the slogan "scientifically-based research" in the United States (Eisenhart & Towne, 2003).

Much of the research synthesized in this book was conducted by researchers who felt more comfortable in the role of curious empathetic collaborator (the interpretive paradigm), a notable trend in science education research over the past three decades (White, 2001). Interpretive research has provided rich descriptions and insightful findings related to traditional and humanistic school science. Its methodologies include ethnography, case studies, R&D, developmental research, and action research (Chapter 4), each with many variations. For instance, in addition to their normal genre, case studies have been written as narratives (Clandinin & Connelly, 1996) and in "sonata form" (Chang & Rosiek, 2003). A completely different methodology, "phenomenography," was introduced to science educators by Erickson (2000). At the center of phenomenological research is a commitment to understanding how students experience the world and how they learn to act in the world. Individual students are not categorized, but instead their relationship to their immediate setting is clarified. Affective and cognitive components merge. This phenomenological approach has yet to be used in a science education study, but it offers a promising avenue for future research, particularly for culture studies in science education.

If science education researchers expect to achieve significant change in school culture and classroom practice, they will perhaps need to be passionate liberators (the critical-theoretic paradigm) generating emancipatory knowledge/practice in the face of seemingly unchangeable organizational structures, relationships, and social conditions (Calabrese Barton & Yang, 2000; Hines, 2003; Pedretti & Hodson, 1995; Roth & Désautels, 2002). The critical-theoretic research paradigm, based in part on poststructuralism, gives emphasis to educational outcomes (e.g., increased equity, social justice, and sense of place) different from interpretive research studies (Calabrese Barton, 2002). Some critical-theoretic methodologies are shared with the interpretive paradigm (Eisenhart, 2001; Kincheloe, 1998). Action research, for example, can integrate educationally sound knowledge with strategic politicization to create classroom change toward a more humanistic science curriculum (Hodson et al., 2002; Tal et al., 2001). Although studies support the effectiveness of action research as a methodology, action research has been limited by its scale and transferability because it usually involves very few unique teachers (Jenkins, 2004; Solomon, 1999b). Critical-theoretic research methodologies also include autoethnography, critical pedagogy, and semiotics (Calabrese Barton, 2001b; Kincheloe, 1998; Taylor, 2004). Apple (1998) warned science educators to draw upon poststructuralist theories in a highly critical manner to avoid transferring them superficially into science education.

Research methodologies have evolved dramatically since the 1960s. Increasingly, teachers and now students tend to be collaborators in the development of curriculum policy, classroom materials, teaching strategies, and even research agendas, along with stakeholders in the local community.

Extensive collaboration in community-based participatory research is a typical feature of the fifth paradigm of educational research: the emerging Aboriginal paradigm. It singularly promotes postcolonial needs and aspirations of Aboriginal nations and communities (Bishop & Glynn, 1999; Menzies et al., 2004; Pillay, 1996; Smith, 1999; Wilson, 2003), summarized in Chapters 2 and 7 (e.g., healing, self-determination, language, and cultural identities). An Aboriginal research paradigm shares values and methods with the critical-theoretic paradigm, particularly the transformational character of its research. However, an Aboriginal paradigm gives particular emphasis to both the "relations between dominant and subordinate groups marked by histories of oppression" (McKinley, 2001, p. 75) and the Aboriginal worldview found in the community engaged in the research. Transformation, holism, spirituality, and relationships are central to this paradigm. For instance, the validity of its research is assessed in part by the quality of the personal relationships forged or strengthened during the study. The reader may wonder if only an Aboriginal person could possibly qualify as a legitimate partner in an Aboriginal research project. Although this is not necessarily the case, the question insinuates enormous challenges and responsibilities a Euro-American faces when conducting research in an Aboriginal setting (Bishop & Glynn, 1999; Menzies et al., 2004).

Which of these paradigms is most valid for advancing humanistic perspectives in school science? White (2001), among others, advocates a thoughtful combination of paradigms in a research program, particularly the quantitative, interpretive, and critical-theoretic paradigms. Longitudinal studies have proven to be a productive research design (Frederick, 1991; Lindahl, 2003; Pedretti et al., 2004; Reiss, 2000; Shapiro, 2004), and they are highly amenable to integrating several paradigms.

CONCLUSION

Western science and technology exert overwhelming power in contemporary globalization, supported by technocratic rationality and free-enterprise philosophies (McKinley et al., 2004). Western science has done this ever since its 17th-century institutionalization as Natural Philosophy at a time of unprecedented colonization (Mendelsohn & Elkana, 1981). Unlike its powerful puzzle-solving counterpart, educational research cannot by itself solve problems for science education. It can, however, increase our community's problem-solving capacity by "sharpening our thinking, directing attention to issues, and encouraging debate" (Fensham, 2004b, p. 13). It is my hope that this book has in some measure added to the problem-solving capacity of our international science education community by reconsidering the purpose of school science in light of the best empirical evidence available and by invigorating the impulse to humanize school science in the context of students' everyday lives.

Without a doubt, science teachers' highest priority is doing what is right for students. This book has explored the many educational and political meanings of "what is right for students." Humanistic researchers are well advised to continue to champion what is right for students. It is a potent force in renegotiating the culture of school science.

My book has highlighted the work of many colleagues who inspire us to follow them, not in their footsteps necessarily, but along a personal path that moves humanistic school science significantly toward science education for everyday life, predicated on evidence-based practice.

References

Abd-El-Khalick, F., Bell, R. L., & Lederman, N. G. (1998). The nature of science and instructional practice: Making the unnatural natural. *Science Education, 82,* 417–436.

Abd-El-Khalick, F., & BouJaoude, S. (1997). An exploratory study of the knowledge base for science teaching. *Journal of Research in Science Teaching, 34,* 673–699.

Abd-El-Khalick, F., & BouJaoude, S. (2004, April). *Do students' views of nature of science influence their conceptual understandings of science content?* Paper presented at the annual meeting of the National Association for Research in Science Teaching, Vancouver, British Columbia, Canada.

Abd-El-Khalick, F., & Lederman, N. G. (2000). Improving science teachers' conceptions of nature of science: A critical review of the literature. *International Journal of Science Education, 22,* 665–701.

Aikenhead, G. S. (1973). The measurement of high school students' knowledge about science and scientists. *Science Education, 51,* 539–549.

Aikenhead, G. S. (1974). Course evaluation. II. Interpretation of student performance on evaluative tests. *Journal of Research in Science Teaching, 11,* 23–30.

Aikenhead, G. S. (1980). *Science in social issues: Implications for teaching.* Ottawa, Ontario, Canada: Science Council of Canada.

Aikenhead, G. S. (1984). Teacher decision making: The case of Prairie High. *Journal of Research in Science Teaching, 21,* 167–186.

Aikenhead, G. S. (1985). Collective decision making in the social context of science. *Science Education, 69,* 453–475.

Aikenhead, G. S. (1988). An analysis of four ways of assessing student beliefs about STS topics. *Journal of Research in Science Teaching, 25,* 607–629.

Aikenhead, G. S. (1989). Decision-making theories as tools for interpreting student behavior during a scientific inquiry simulation. *Journal of Research in Science Teaching, 26,* 189–203.

Aikenhead, G. S. (1992). Logical reasoning in science and technology. *Bulletin of Science, Technology & Society, 12,* 149–159.

Aikenhead, G. S. (1994a). Collaborative research and development to produce an STS course for school science. In J. Solomon & G. Aikenhead (Eds.), *STS education: International perspectives on reform* (pp. 216–227). New York: Teachers College Press.

Aikenhead, G. S. (1994b). Consequences to learning science through STS: A research perspective. In J. Solomon & G. Aikenhead (Eds.), *STS education: International perspectives on reform* (pp. 169–186). New York: Teachers College Press.

Aikenhead, G. S. (1994c). The social contract of science: Implications for teaching science. In J. Solomon & G. Aikenhead (Eds.), *STS education: International perspectives on reform* (pp. 11–20). New York: Teachers College Press.

Aikenhead, G. S. (1994d). What is STS teaching? In J. Solomon & G. Aikenhead (Eds.), *STS education: International perspectives on reform* (pp. 47–59). New York: Teachers College Press.

Aikenhead, G. S. (1996). Science education: Border crossing into the subculture of science. *Studies in Science Education, 27,* 1–51.

Aikenhead, G. S. (1997). Toward a First Nations cross-cultural science and technology curriculum. *Science Education, 81,* 217–238.

Aikenhead, G. S. (2000a). *Rekindling Traditions: Cross-cultural science & technology units.* Retrieved December 1, 2004, from http://capes.usask.ca/ccstu.

Aikenhead, G. S. (2000b). Renegotiating the culture of school science. In R. Millar, J. Leach, & J. Osborne (Eds.), *Improving science education: The contribution of research* (pp. 245–264). Birmingham, UK: Open University Press.

Aikenhead, G. S. (2000c). STS science in Canada: From policy to student evaluation. In D. D. Kumar & D. E. Chubin (Eds.), *Science, technology, and society: A sourcebook on research and practice* (pp. 49–89). New York: Kluwer Academic/Plenum.

Aikenhead, G. S. (2001a). Integrating Western and Aboriginal sciences: Cross-cultural science teaching. *Research in Science Education, 31,* 337–335.

Aikenhead, G. S. (2001b). Students' ease in crossing cultural borders into school science. *Science Education, 85,* 180–188.

Aikenhead, G. S. (2002a). Cross-cultural science teaching: *Rekindling Traditions* for Aboriginal students. *Canadian Journal of Science, Mathematics and Technology Education, 2,* 287–304.

Aikenhead, G.S. (2002b). The educo-politics of curriculum development. *Canadian Journal of Science, Mathematics and Technology Education, 2,* 49–57.

Aikenhead, G. S. (2003). STS education: A rose by any other name. In R. Cross (Ed.), *A vision for science education: Responding to the work of Peter Fensham* (pp. 59–75). New York: RoutledgeFalmer.

Aikenhead, G. S. (2005). Science-based occupations and the science curriculum: Concepts of evidence. *Science Education, 89,* 242–275.

Aikenhead, G. S., Fleming, R. W., & Ryan, A. G. (1987). High school graduates' beliefs about science-technology-society. Part I. Methods and issues in monitoring students' views. *Science Education, 71,* 145–161.

Aikenhead, G. S., & Huntley, B. (1999). Teachers' views on Aboriginal students learning Western and Aboriginal science. *Canadian Journal for Native Education, 23,* 159–175.

Aikenhead, G. S., & Jegede, O. J. (1999). Cross-cultural science education: A cognitive explanation of a cultural phenomenon. *Journal of Research in Science, 36,* 269–287.

Aikenhead, G. S., & Otsuji, H. (2000). Japanese and Canadian science teachers' views on science and culture. *Journal of Science Teacher Education, 11*, 277–299.

Aikenhead, G. S., & Ryan, A. G. (1992). The development of a new instrument: "Views on science-technology-society" (VOSTS). *Science Education, 76*, 477–491.

Aikenhead, G. S., Ryan, A. G., & Fleming, R. W. (1989). *Views on science-technology-society*. Saskatoon, Canada: University of Saskatchewan, Department of Curriculum Studies.

Alaska Native Knowledge Network [ANKN]. (1998). *Alaska standards for culturally-responsive schools*. Fairbanks, AK: University of Alaska.

Alaska Native Knowledge Network [ANKN]. (2003). *Guidelines for cross-cultural orientation programs*. Fairbanks, AK: University of Alaska.

Alaska Native Knowledge Network [ANKN]. (2004a). *Alaska Native Knowledge Network*. Retrieved November 24, 2004, from http://www.ankn.uaf.edu/.

Alaska Native Knowledge Network [ANKN]. (2004b). *Spiral pathway for integrating rural Alaska learning*. Retrieved November 24, 2004, from http://www.ankn.uaf.edu/.

Allchin, D. (2003). Scientific myth-conceptions. *Science Education, 87*, 329–351.

Allchin, D. (2004). Should the sociology of science be rated X? *Science Education, 88*, 934–946.

Allen, N. J., & Crawley, F. E. (1998). Voices from the bridge: Worldview conflicts of Kickapoo students of science. *Journal of Research in Science Teaching, 35*, 111–132.

Alsop, S. (2003). Pupils, science, research, practice, and politics: Musings on the emergence of a pre-paradigmatic field. *Canadian Journal of Science, Mathematics and Technology Education, 3*, 281–285.

Alters, B. J. (1997). Whose nature of science? *Journal of Research in Science Teaching, 34*, 39–55.

American Association for the Advancement of Science. [AAAS]. (1977). *Native Americans in science*. Washington, DC: Author.

American Association for the Advancement of Science. [AAAS]. (1989). *Project 2061: Science for all Americans*. Washington, DC: Author.

American Association for the Advancement of Science. [AAAS]. (2002). *150 years of advancing science: A history of AAAS*. Retrieved December 15, 2003, from http://archives.aaas.org/exhibit/origins.

Anamuah-Mensah, J. (1999). The science and technology in action in Ghana (STAG): Anatomy of a process. In M. Keogh & S. Naidoo (Eds.), *Proceedings of the 9th symposium of IOSTE*: Vol. 1, pp. 31–35. Durban, South Africa: University of Durban-Westville.

Anderson, R. D., & Helms, J. V. (2001). The ideal of standards and the reality of schools: Needed research. *Journal of Research in Science Teaching, 38*, 3–16.

Anderson, T., & Kilbourn, B. (1983). Creation, evolution, and curriculum. *Science Education, 67*, 45–55.

Apple, M. (1996). *Cultural politics and education*. New York: Teachers College Press.

Apple, M. (1998). Mechanisms of differentiation: A response to Roth and McGinn. *Journal of Research in Science Teaching, 35*, 423–425.

Atkin, M., & Helms, J. (1993). Getting serious about priorities in science education. *Studies in Science Education, 21*, 1–20.

Atwater, M. M. (1996). Social constructivism: Infusion into the multicultural science education research agenda. *Journal of Research in Science Teaching, 33*, 821–837.

Atwater, M. M., & Crockett, D. (2003). Prospective teachers' education world view and teacher education programs: Through the eyes of culture, ethnicity, and class. In S. M. Hines (Ed.), *Multicultural science education: Theory, practice, and promise* (pp. 55–86). New York: Peter Lang.

Bajracharya, H., & Brouwer, W. (1997). A narrative approach to science teaching in Nepal. *International Journal of Science Education, 19*, 429–446.

Baker, D., & Taylor, P. C. S. (1995). The effect of culture on the learning of science in non-western countries: The results of an integrated research review. *International Journal of Science Education, 17*, 695–704.

Barker, M. (2004, July). *Indigenous cultures and the culture of science: The dilemma of being and knowing.* Paper presented to the 11th International Organization for Science & Technology Education (IOSTE) symposium, Lublin, Poland.

Barnett, J., & Hodson, D. (2001). Pedagogical context knowledge: Toward a fuller understanding of what good science teachers know. *Science Education, 85*, 426–453.

Barnhardt, R., Kawagley, A. O., & Hill, F. (2000). Cultural standards and test scores. *Sharing Our Pathways, 5*(4), 1–4.

Barrett, S. E., & Pedretti, E. (2004, April). *Contrasting orientations to science-technology-society-environment education.* Paper presented at the annual meeting of the National Association for Research in Science Teaching, Vancouver, British Columbia, Canada.

Bartholomew, H., Osborne, J., & Ratcliffe, M. (2004). Teaching students "ideas-about-science": Five dimensions of effective practice. *Science Education, 88*, 655–682.

Battiste, M. (Ed.). (2000). *Reclaiming Indigenous voice and vision.* Vancouver, Canada: University of British Columbia Press.

Battiste, M., & Barman, J. (Eds.). (1995). *First Nations education in Canada: The circle unfolds.* Vancouver, Canada: University of British Columbia Press.

Battiste, M., & Henderson, J. Y. (2000). *Protecting Indigenous knowledge and heritage.* Saskatoon, Saskatchewan, Canada: Purich.

Becher, A. (2002, Winter). Native woman earns recognition as an American environmental leader. *Winds of Change, 17*, 60–76.

Bell, R. L., & Lederman, N. G. (2003). Understandings of the nature of science and decision making on science and technology based issues. *Science Education, 87*, 352–377.

Bencze, L., & Hodson, D. (1999). Changing practice by changing practice: Toward more authentic science and science curriculum development. *Journal of Research in Science Teaching, 36*, 521–539.

Benjamin, H. (1939). *The saber-tooth curriculum.* New York: McGraw-Hill.

Berkowitz, P. (2001, December). Western science meets Mi'kmaq knowledge. *University Affairs, 42*, 16–21.

Bianchini, J. A., Cavazos, L. M., & Helm, J. V. (2000). From professional lives to inclusive practice: Science teachers and scientists' views of gender and ethnicity in science education. *Journal of Research in Science Teaching, 37*, 511–547.

Bianchini, J. A., Johnston, C. C., Oram, S. Y., & Cavazos, L. M. (2003). Learning to teach science in contemporary and equitable ways: The successes and struggles of first-year science teachers. *Science Education, 87*, 419–443.

Bianchini, J. A., & Solomon, E. M. (2003). Constructing views of science tied to issues of equity and diversity: A study of beginning science teachers. *Journal of Research in Science Teaching, 40*, 53–76.

Bingle, W. H., & Gaskell, P. J. (1994). Scientific literacy for decisionmaking and the social construction of scientific knowledge. *Science Education, 78*, 185–201.

Bishop, J. M. (1995). Enemies of promise. *The Wilson Quarterly, 19*(3), 61–65.

Bishop, R., & Glynn, T. (1999). *Culture counts: Changing power relations in education*. Palmerston North, New Zealand: Dunmore Press.

Black, P. (1995). Curriculum and assessment in science education: The policy interface. *International Journal of Science Education, 17*, 453–469.

Black, P. (1998). Assessment by teachers and the improvement of students' learning. In B. J. Fraser & K. G. Tobin (Eds.), *International handbook of science education* (pp. 811–822). Dordrecht, The Netherlands: Kluwer Academic.

Blades, D. (1997). *Procedures of power and curriculum change*. New York: Peter Lang.

Botton, C., & Brown, C. (1998). The reliability of some VOSTS items when used with preservice secondary science teachers in England. *Journal of Research in Science Teaching, 35*, 53–71.

Bouillion, L. M., & Gomez, L. M. (2001). Connecting school and community with science learning. *Journal of Research in Science Teaching, 38*, 878–898.

Bourdieu, P., & Passeron, J. (1977). *Reproduction in education, society and culture*. London: Sage.

Boydston, T. L. (1999). Policy-making in science systemic reform. *Research in Science Education, 29*, 141–157.

Brand, B. R., Glasson, G. E., & Green, A. M. (2004, April). *Perspectives of female and minority students on science learning*. Paper presented at the annual meeting of the National Association for Research in Science Teaching, Vancouver, British Columbia, Canada.

Brandt, C. B. (2004, April). *To walk in beauty: Performance, gender, and discursive space in Western science*. Paper presented at the annual meeting of the National Association for Research in Science Teaching, Vancouver, British Columbia, Canada.

Brickhouse, N. W. (1989). The teaching of philosophy of science in secondary classrooms: Case study of teachers' personal theories. *International Journal of Science Education, 11*, 437–449.

Brickhouse, N. W. (2001). Embodying science: A feminist perspective on learning. *Journal of Research in Science Teaching, 38*, 282–295.

Brickhouse, N. W. (2003). Science for all? Science for girls? Which girls? In R. Cross (Ed.), *A vision for science education: Responding to the work of Peter Fensham* (pp. 93–101). New York: RoutledgeFalmer.

Brickhouse, N. W., Lowery, P., & Schultz, K. (2000). What kind of a girl does science: The construction of school science identities. *Journal of Research in Science Teaching, 37,* 441–458.

Brickhouse, N. W., & Potter, J. T. (2001). Young women's scientific identity formation in an urban context. *Journal of Research in Science Teaching, 38,* 965–980.

Briscoe, C. (1991). The dynamic interactions among beliefs, role metaphors, and teaching practices: A case study of teacher change. *Science Education, 75,* 185–199.

Brown, B. A. (2004). Discourse identity: Assimilation into the culture of science and its implications for minority students. *Journal of Research in Science Teaching, 41,* 810–834.

Brown, B. A., Reveles, J. M., & Kelly, G. (2004, April). *Scientific literacy and discursive identity: A theoretical framework for understanding science education.* Paper presented at the annual meeting of the National Association for Research in Science Teaching, Vancouver, British Columbia, Canada.

Brush, L. (1979). Avoidance of science and stereotypes of scientists. *Journal of Research in Science Teaching, 16,* 237–241.

Bybee, R. W. (1993). *Reforming science education.* New York: Teachers College Press.

Bybee, R. W., & Bonnstetter, R. J. (1987). What research says: Implementing the science-technology-society theme in science education: Perceptions of science teachers. *School Science and Mathematics, 87,* 144–152.

Byrne, M. S., & Johnstone, A. H. (1988). How to make science relevant. *School Science Review, 70*(251), 325–339.

Cajas, F. (1998). Using out-of-school experience in science lessons: An impossible task? *International Journal of Science Education, 20,* 623–625.

Cajas, F. (1999). Public understanding of science: Using technology to enhance school science in everyday life. *International Journal of Science Education, 21,* 765–773.

Cajete, G. A. (1986). *Science: A Native American perspective.* Unpublished doctoral dissertation, International College, Los Angeles, CA.

Cajete, G. A. (1999). *Igniting the sparkle: An Indigenous science education model.* Skyand, NC: Kivaki Press.

Cajete, G. (2000a). Indigenous knowledge: The Pueblo metaphor of Indigenous education. In M. Battiste (Ed.), *Reclaiming Indigenous voice and vision* (pp. 181–191). Vancouver, Canada: University of British Columbia Press.

Cajete, G. A. (2000b). *Native science: Natural laws of interdependence.* Santa Fe, NM: Clear Light.

Calabrese Barton, A. (1998). Teaching science with homeless children: Pedagogy, representation, and identity. *Journal of Research in Science Teaching, 35,* 379–394.

Calabrese Barton, A. (2001a). Capitalism, critical pedagogy, and urban science education: An interview with Peter McLaren. *Journal of Research in Science Teaching, 38,* 847–859.

Calabrese Barton, A. (2001b). Science education in urban settings: Seeking new ways of praxis through critical ethnography. *Journal of Research in Science Teaching, 38,* 899–917.

Calabrese Barton, A. (2002). Urban science education studies: A commitment to equity, social justice and a sense of place. *Studies in Science Education, 38*, 1–38.

Calabrese Barton, A. (2004). *Teaching science for social justice*. New York: Teachers College Press.

Calabrese Barton, A., & Osborne, M. D. (2001). *Teaching science in diverse settings: Marginalized discourses and classroom practice*. New York: Peter Lang.

Calabrese Barton, A., & Yang, K. (2000). The case of Miguel and the culture of power in science. *Journal of Research in Science Teaching, 37*, 871–889.

Campbell, B., Lazonby, J., Millar, R., Nicolson, P., Ramsden, J., & Waddington, D. (1994). Science: The Salters' approach—A case study of the process of large scale curriculum development. *Science Education, 78*, 415–447.

Campbell, B., & Lubben, F. (2000). Learning science through contexts: Helping pupils make sense of everyday situations. *International Journal of Science Education, 22*, 239–252.

Carlone, H. B. (2003). Innovative science within and against a culture of "achievement." *Science Education, 87*, 307–328.

Carlone, H. B. (2004). The cultural production of science in reform-based physics: Girls' access, participation and resistance. *Journal of Research in Science Teaching, 41*, 392–414.

Carlsen, W., Kelly, G., & Cunningham, C. (1994). Teaching *ChemCom*: Can we use the text without being used by the text? In J. Solomon & G. Aikenhead (Eds.), *STS education: International perspectives on reform* (pp. 84–96). New York: Teachers College Press.

Casteel, J. D., & Yager, R. E. (1968). The University of Iowa science and culture project. *School Science and Mathematics, 68*, 412–416.

Champagne, A. B., & Klopfer, L. E. (1982). A causal model of students' achievement in a college physics course. *Journal of Research in Science Teaching, 19*, 299–309.

Chang, P. J., & Rosiek, J. (2003). Anti-colonialist antinomies in a biology lesson: A sonata-form case study of cultural conflict in a science classroom. *Curriculum Inquiry, 33*, 251–290.

Cheek, D. W. (1992). *Thinking constructively about science, technology, and society education*. Albany: State University of New York Press.

Cheek, D. W. (2000). Marginalization of technology within the STS movement in American K–12 education. In D. D. Kumar & D. E. Chubin (Eds.), *Science, technology, and society: A sourcebook on research and practice* (pp. 167–192). New York: Kluwer Academic/Plenum.

Chiang-Soong, B., & Yager, R. E. (1993). The inclusion of STS material in the most frequently used secondary science textbooks in the U.S. *Journal of Research in Science Teaching, 30*, 339–349.

Chin, P., Munby, H., Hutchinson, N. L., Taylor, J., & Clark, F. (2004). Where's the science? Understanding the form and function of workplace science. In E. Scanlon, P. Murphy, J. Thomas, & E. Whitelegg (Eds.), *Reconsidering science learning* (pp. 118–134). New York: RoutledgeFalmer.

Chinn, P. W. U. (1999). Multiple words/mismatched meanings: Barriers to minority women engineers. *Journal of Research in Science Teaching, 36*, 621–636.

Chinn, P. W. U. (2004, July). *Moving Native Hawaiian knowledge from the margins to the center: The transformative potential of place and standards-based, culture-science professional development*. Paper presented to the Culture Studies in Science Education (CSSE) meeting, Kobe University, Japan.

Cho, F. (2002). The development of an alternative in-service programme for Korean science teachers with an emphasis on science-technology-society. *International Journal of Science Education, 24*, 1021–1035.

Christie, M. J. (1991). Aboriginal science for the ecologically sustainable future. *Australian Science Teachers Journal, 37*(1), 26–31.

Clagett, M. (1963). *Greek science in antiquity*. New York: Collier Books.

Clandinin, D. J., & Connelly, F. M. (1996). Teachers' professional knowledge landscapes: Teacher stories—stories of teachers—school stories—stories of schools. *Educational Research, 25*, 24–30.

Cleaves, A. (2005). The formation of science choices in secondary school. *International Journal of Science Education, 27*, 471–486.

Cobern, W. W. (1991). *World view theory and science education research* (Monograph Series, Vol. 3). Cincinnati, OH: National Association for Research in Science Teaching.

Cobern, W. W. (1993). College students' conceptualizations of nature: An interpretive world view analysis. *Journal of Research in Science Teaching, 30*, 935–951.

Cobern, W. W. (1996a). Constructivism and non-Western science education research. *International Journal of Science Education, 18*, 295–310.

Cobern, W. W. (1996b). Worldview theory and conceptual change in science education. *Science Education, 80*, 579–610.

Cobern, W. W. (2000). *Everyday thoughts about nature*. Boston: Kluwer Academic.

Cobern, W. W., & Aikenhead, G. S. (1998). Cultural aspects of learning science. In B. J. Fraser & K. G. Tobin (Eds.), *International handbook of science education* (pp. 39–52). Dordrecht, The Netherlands: Kluwer Academic.

Cobern, W. W., Gibson, A. T., & Underwood, S. A. (1999). Conceptualizations of nature: An interpretive study of 16 ninth graders' everyday thinking. *Journal of Research in Science Teaching, 36*, 541–564.

Coburn, C. E. (2003). Rethinking scale: Moving beyond numbers to deep and lasting change. *Educational Researcher, 32*(6), 3–12.

Cole, S. (1992). *Making science: Between nature and society*. Cambridge, MA: Harvard University Press.

Coles, M. (1996). What does industry want from science education? In K. Calhoun, R. Panwar, & S. Shrum (Eds.), *Proceedings of the 8th symposium of IOSTE*: Vol. 1, pp. 292–300. Edmonton, Canada: University of Alberta, Faculty of Education.

Coles, M. (1998). Science for employment and higher education. *International Journal of Science Education, 20*, 609–621.

Collingridge, D. (1989). Incremental decision making in technological innovations: What role for science? *Science, Technology, & Human Values, 14*, 141–162.

Collins, A. (2004). Guest editorial. *Science Education, 88*, 1–3.

Contreras, A., & Lee, O. (1990). Differential treatment of students by middle school science teachers: Unintended cultural bias. *Science Education, 74*, 433–444.

Costa, V. B. (1995). When science is "another world": Relationships between worlds of family, friends, school, and science. *Science Education, 79*, 313–333.

Costa, V. (1997a). Honours chemistry: High-status knowledge or knowledge about high status? *Journal of Curriculum Studies, 29*, 289–313.

Costa, V. (1997b). How teacher and students study "all that matters" in high school chemistry. *International Journal of Science Education, 19*, 1005–1023.

Council of Ministers of Education of Canada. (1997). *Common framework of science learning outcomes*. Ottawa, Ontario, Canada: Author.

Council of Science and Technology Institutes. (1993). *Mapping the science, technology and mathematics domain*. London: Author.

Cronin-Jones, L. (1991). Science teacher beliefs and their influence on curriculum implementation: Two case studies. *Journal of Research in Science Teaching, 28*, 235–250.

Cross, R. T. (1997a). Ideology and science teaching: Teachers' discourse. *International Journal of Science Education, 19*, 607–616.

Cross, R. T. (1997b). The schooling of science as a deterrent to democracy in Australia: The example of benzene. *International Journal of Science Education, 19*, 469–479.

Cross, R. T., & Ormiston-Smith, H. (1996). Physics teaching, professional development and a socially critical ideology. *Journal of Curriculum Studies, 28*, 651–667.

Cross, R. T., & Price, R. F. (1992). *Teaching science for social responsibility*. Sydney: St. Louis Press.

Cross, R. T., & Price, R. F. (1996). Science teachers' social conscience and the role of controversial issues in the teaching of science. *Journal of Research in Science Teaching, 33*, 319–333.

Cross, R. T., & Price, R. F. (1999). The social responsibility of science and public understanding. *International Journal of Science Education, 21*, 775–785.

Cross, R. T., & Price, R. F. (2002). Teaching controversial science for social responsibility: The case of food production. In W-M. Roth & J. Désautels (Eds.), *Science education as/for sociopolitical action* (pp. 99–123). New York: Peter Lang.

Cuban, L. (1994, June 15). The great school scam. *Education Week*, p. 44.

Cunningham, C. M. (1998). The effect of teachers' sociological understanding of science (SUS) on curricular innovation. *Research in Science Education, 28*, 243–257.

Cutcliffe, S. H. (1989). The emergence of STS as an academic field. In P. Durbin (Ed.), *Research in philosophy and technology* (Vol. 9, pp. 287–301). Greenwich, CT: JAI Press.

Dahncke, H. (1996). Science learning and competence to act: A report on a study on computer simulations and STS teaching. In K. Calhoun, R. Panwar, & S. Shrum (Eds.), *Proceedings of the 8th symposium of IOSTE*: Vol. 1, pp. 24–31. Edmonton, Canada: University of Alberta, Faculty of Education.

Dampier, W. C. (1971). *A history of science* (4th ed.). Cambridge: Cambridge University Press.

David, E. E. (1995). A realistic scenario for U.S. R&D. *Bulletin of Science, Technology & Society, 15*, 14–18.

David, M. R. (2003). *Biographical influences on teaching STSE: A case study*. Unpublished master's thesis, University of Saskatchewan, Saskatoon, Canada.

Davies, I. (2004). Science and citizenship education. *International Journal of Science Education, 26*, 1751–1763.

Davis, K. S. (2003). "Change is hard": What science teachers are telling us about reform and teacher learning of innovative practices. *Science Education, 87*, 3–30.

Dawson, V. (2004, April). *High school students' ethical decision-making in a forensic DNA testing context*. Paper presented at the annual meeting of the National Association for Research in Science Teaching, Vancouver, British Columbia, Canada.

Dawson, V., & Taylor, P. C. (2000). Do adolescents' bioethical decisions differ from those of experts? *Journal of Biological Education, 34*(4), 1–5.

DeBoer, G. E. (1991). *A history of ideas in science education*. New York: Teachers College Press.

Dekkers, J., & Delaeter, J. (2001). Enrollment trends in school science education in Australia. *International Journal of Science Education, 23*, 487–500.

Deloria, V. (1992, Autumn). Relativity, relatedness and reality. *Winds of Change, 7*, 35–40.

Del Giorno, B. J. (1969). The impact of changing scientific knowledge on science education in the United States since 1850. *Science Education, 53*, 191–195.

DeMerchant, R. V. (2002). *A case study of integrating Inuuqatigiit into a Nunavut junior high school classroom*. Unpublished master's thesis, University of Saskatchewan, Saskatoon, Canada.

Désautels, J., Fleury, S. C., & Garrison, J. (2002). The enactment of epistemological practice as subversive social action, the provocation of power, and anti-modernism. In W-M. Roth & J. Désautels (Eds.), *Science education as/for sociopolitical action* (pp. 237–269). New York: Peter Lang.

De Vos, W., & Reiding, J. (1999). Public understanding of science as a separate subject in secondary schools in The Netherlands. *International Journal of Science Education, 21*, 711–719.

Dewey, J. (1916). *Democracy and education*. New York: Macmillan.

Dhingra, K. (2003). Thinking about television science: How students understand the nature of science from different program genres. *Journal of Research in Science Teaching, 40*, 234–256.

Dierking, L. D., Ellenbogen, K. M., & Falk, J. H. (2004). In principle, in practice: Perspectives on a decade of museum learning research (1994–2004). *Science Education, 88*(Suppl. 1), S1–S3.

Dimopoulos, K., & Koulaidis, V. (2003). Science and technology education for citizenship: The potential role of the press. *Science Education, 87*, 241–256.

Donnelly, J. F. (2002). The "humanist" critique of the place of science in the curriculum in the nineteenth century, and its continuing legacy. *History of Education, 31*, 535–555.

Donnelly, J. F. (2004). Humanizing science education. *Science Education, 88*, 762–784.

Dori, Y. J., & Tal, R. T. (2000). Formal and informal collaborative projects: Engaging in industry with environmental awareness. *Science Education, 84*, 95–113.

Dreyer, J. L. E. (1953). *A history of astronomy from Thales to Kepler* (2nd ed.). New York: Dover.

Driver, R. (1989). Students' conceptions and the learning of science. *International Journal of Science Education, 11*, 481–490.

Driver, R., Asoko, H., Leach, J., Mortimer, E., & Scott, P. (1994). Constructing scientific knowledge in the classroom. *Educational Researcher, 23*, 5–12.

Driver, R., Leach, J., Millar, R., & Scott, P. (1996). *Young people's images of science*. Buckingham, UK: Open University Press.

Driver, R., Newton, P., & Osborne, J. (2000). Establishing the norms of scientific argumentation in classrooms. *Science Education, 84*, 287–312.

Drori, G. S. (1998). A critical appraisal of science education for economic development. In W. W. Cobern (Ed.), *Socio-cultural perspectives on science education* (pp. 49–74). Boston: Kluwer Academic.

Duffee, L., & Aikenhead, G. S. (1992). Curriculum change, student evaluation, and teacher practical knowledge. *Science Education, 76*, 493–506.

Duggan, S., & Gott, R. (2002). What sort of science education do we really need? *International Journal of Science Education, 24*, 661–679.

Duschl, R. A. (1988). Abandoning the scientistic legacy of science education. *Science Education, 72*, 51–62.

Duschl, R. A., & Gitomer, D. (1996, April). *Project Sepia design principles*. Paper presented at the annual meeting of the American Educational Research Association, New York.

Dzama, E. N. N., & Osborne, J. F. (1999). Poor performance in science among African students: An alternative explanation to the African worldview thesis. *Journal of Research in Science Teaching, 36*, 387–405.

Eijkelhof, H. M. C. (1990). *Radiation and risk in physics education*. Utrecht, The Netherlands: University of Utrecht CDβ Press.

Eijkelhof, H. M. C. (1994). Toward a research base for teaching ionizing radiation in a risk perspective. In J. Solomon & G. Aikenhead (Eds.), *STS education: International perspectives on reform* (pp. 205–215). New York: Teachers College Press.

Eijkelhof, H. M. C., & Kapteijn, M. (2000). A new course on public understanding of science for senior general secondary education in the Netherlands. In R. T. Cross & P. J. Fensham (Eds.), *Science and the citizen for educators and the public* (pp. 189–199). Melbourne: Arena.

Eijkelhof, H. M. C., Klaassen, C. W. J. M., Lijnse, P. L., & Scholte, R. L. J. (1990). Perceived incidence and importance of lay-idea on ionizing radiation: Results of a Delphi-study among radiation experts. *Science Education, 74*, 183–195.

Eijkelhof, H. M. C., & Kortland, K. (1987). Physics in its personal social and scientific context. *Bulletin of Science, Technology & Society, 7*, 125–136.

Eijkelhof, H. M. C., & Lijnse, P. (1988). The role of research and development to improve STS education: Experiences from the PLON project. *International Journal of Science Education, 10*, 464–474.

Eisenhart, M. (2001). Changing conceptions of culture and ethnographic method-
ology: Recent thematic shifts and their implications for research on teaching.
In V. Richardson (Ed.), *Handbook of research on teaching* (4th ed. pp. 209–225).
Washington, DC: American Educational Research Association.

Eisenhart, M., Finkel, E., & Marion, S. (1996). Creating the conditions for scientific literacy: A re-examination. *American Educational Research Journal, 33,* 261–
295.

Eisenhart, M., & Towne, L. (2003). Contestation and change in national policy
on "scientifically based" education research. *Educational Researcher, 32*(7), 31–
38.

Eisenkraft, A. (1998). *Active physics.* Armonk, NY: It's About Time.

Elmore, R. F. (2003, March). *Large-scale improvement of teaching and learning: What
we know, what we need to know.* Paper presented at the annual meeting of the
National Association for Research in Science Teaching, Philadelphia, PA.

Erickson, F. (2004). Culture in society and educational practices. In J. A. Banks
& C. A. M. Banks (Eds.), *Multicultural education: Issues and perspectives* (5th ed.,
pp. 31–60). New York: John Wiley.

Erickson, G. (2000). Research programmes and the student science learning literature. In R. Millar, J. Leach, & J. Osborne (Eds.), *Improving science education: The contribution of research* (pp. 271–292). Birmingham, UK: Open
University Press.

Ezeife, A. N. (2003). The pervading influence of cultural border crossing and collateral learning on the learner of science and mathematics. *Canadian Journal
of Native Education, 27,* 179–194.

Fensham, P. J. (1988). Approaches to the teaching of STS in science education.
International Journal of Science Education, 10, 346–356.

Fensham, P. J. (1992). Science and technology. In P. W. Jackson (Ed.), *Handbook
of research on curriculum* (pp. 789–829). New York: Macmillan.

Fensham, P. J. (1993). Academic influence on school science curricula. *Journal of
Curriculum Studies, 25,* 53–64.

Fensham, P. J. (1994). STS and comparative assessment of scientific literacy. In
K. Boersma, K. Kortland, & J. van Trommel (Eds.), *7th IOSTE symposium pro-
ceedings: Papers.* Part 1 (pp. 255–262). Endrecht, The Netherlands: IOSTE
Conference Committee.

Fensham, P. J. (1998). The politics of legitimating and marginalizing companion
meanings: Three Australian case stories. In D. A. Roberts & L. Östman (Eds.),
Problems of meaning in science curriculum (pp. 178–192).New York: Teachers
College Press.

Fensham, P. J. (2000a). Issues for schooling in science. In R. T. Cross & P. J.
Fensham (Eds.), *Science and the citizen for educators and the public* (pp. 73–77).
Melbourne: Arena.

Fensham, P. J. (2000b). Providing suitable content in the "science for all" curriculum. In R. Millar, J. Leach, & J. Osborne (Eds.), *Improving science education: The contribution of research* (pp. 147–164). Birmingham, UK: Open
University Press.

Fensham, P. J. (2002). Time to change drivers for scientific literacy. *Canadian Journal of Science, Mathematics and Technology Education, 2,* 9–24.

Fensham, P. J. (2004a, July). *Beyond knowledge: Other outcome qualities for science education.* Keynote presentation to the 11th International Organization for Science & Technology Education (IOSTE) symposium, Lublin, Poland.

Fensham, P. J. (2004b). *Defining an identity: The evolution of science education as a field of research.* Boston: Kluwer Academic.

Fensham P. J., & Corrigan, D. (1994). The implementation of an STS chemistry course in Australia: A research perspective. In J. Solomon & G. Aikenhead (Eds.), *STS education: International perspectives on reform* (pp. 194–204). New York: Teachers College Press.

Fleming, R. (1986a). Adolescent reasoning in socio-scientific issues, Part I: Social cognition. *Journal of Research in Science Teaching, 23,* 677–687.

Fleming, R. (1986b). Adolescent reasoning in socio-scientific issues, Part II: Non-social cognition. *Journal of Research in Science Teaching, 23,* 689–698.

Foucault, M. (1980). *Power/knowledge: Selected interviews and other writings.* New York: Pantheon Books.

Fourez, G. (1989). Scientific literacy, societal choices, and ideologies. In A. B. Champagne, B. E. Lovitts, & B. J. Calinger (Eds.), *Scientific literacy* (pp. 89–108). Washington, DC: American Association for the Advancement of Science.

Fourez, G. (1997). Scientific and technological literacy as social practice. *Social Studies of Science, 27,* 903–936.

Frederick, W. A. (1991). Science and technology education: An engineer's perspective. In S. K. Majumdar, L. M. Rosenfeld, P. A. Rubba, E. W. Miller, & R. F. Schmalz (Eds.), *Science education in the United States: Issues, crises and priorities* (pp. 386–393). Easton, PA: The Pennsylvania Academy of Science.

Fuller, S. (1997). *Science.* Minneapolis: University of Minnesota Press.

Furnham, A. (1992). Lay understanding of science: Young people and adults' ideas of scientific concepts. *Studies in Science Education, 20,* 29–64.

Fusco, D. (2001). Creating relevant science through urban planning and gardening. *Journal of Research in Science Teaching, 38,* 860–877.

Gagné, M-C. (2003, August). *Creative activities as a way of helping more individuals feel comfortable with their identity as science students.* Paper presented at the European Science Education Research Association (ESERA) 2003 conference, Noordwijkerhout, The Netherlands.

Gagné, M-C. (2004, April). *Valuing identity and knowledge: "Rekindling Traditions" as an exemplar for minority education.* Paper presented at the annual meeting of the National Association for Research in Science Teaching, Vancouver, British Columbia, Canada.

Galili, I., & Hazan, A. (2001). The effect of a history-based course in optics on students' views about science. *Science & Education, 10,* 7–32.

Galindo, E., & Barta, J. (2001, Autumn). Indian summer. *Winds of Change, 16,* 54–56.

Gallagher, J. J. (1971). A broader base for science education. *Science Education, 55,* 329–338.

Gallagher, J. J. (1987). A summary of research in science education. *Science Education, 71*, 277–284.

Gallagher, J. J. (1991). Prospective and practicing secondary school science teachers' knowledge and beliefs about the philosophy of science. *Science Education, 75*, 121–133.

Gallagher, J. J. (1998). Science teaching as shared culture: An international perspective. *NARST News, 41*(3), 4.

Gardner, P. L. (1998). Students' interest in science and technology: Gender, age and other factors. In L. Hoffmann, A. Krapp, K. A. Renninger, & J. Baumert (Eds.), *Proceedings of the Seeon conference on interest and gender* (pp. 41–57). Kiel, Germany: IPN, University of Kiel.

Garnett, P. J., & Tobin, K. (1989). Teaching for understanding: Exemplary practice in high school chemistry. *Journal of Research in Science Teaching, 26*, 1–14.

Gaskell, J. P. (1982). Science, technology and society: Issues for science teachers. *Studies in Science Education, 9*, 33–46.

Gaskell, J. P. (1989). Science and technology in British Columbia: A course in search of a community. *Pacific Education, 1*(3), 1–10.

Gaskell, J. P. (1992). Authentic science and school science. *International Journal of Science Education, 14*, 265–272.

Gaskell, J. P. (1994). Assessing STS literacy: What is rational? In K. Boersma, K. Kortland, & J. van Trommel (Eds.), *7th IOSTE symposium proceedings: Papers*. Part 1, pp. 309–320. Endrecht, The Netherlands: IOSTE Conference Committee.

Gaskell, P. J. (2003). Perspectives and possibilities in the politics of science curriculum. In R. Cross (Ed.), *A vision for science education: Responding to the work of Peter Fensham* (pp. 139–152). New York: RoutledgeFalmer.

Gaskell, P. J., & Hepburn, G. (1998). The course as token: A construction of/by networks. *Research in Science Education, 28*, 65–76.

Geddis, A. (1991). Improving the quality of classroom discourse on controversial issues. *Science Education, 75*, 169–183.

Gee, J. P. (2002). Identity as an analytic lens for research in education. *Review of Research in Education, 25*, 99–125.

George, J. M. (1999a). Indigenous knowledge as a component of the school curriculum. In L. M. Semali & J. L. Kincheloe (Eds.), *What is indigenous knowledge? Voices from the academy* (pp. 79–94). New York: Falmer Press.

George, J. M. (1999b). World view analysis of knowledge in a rural village: Implications for science education. *Science Education, 83*, 77–95.

George, J., & Glasgow, J. (1988). Street science and conventional science in the West Indies. *Studies in Science Education, 15*, 109–118.

Gibbs, W.W., & Fox, D. (1999, October). The false crises in science education. *Scientific American*, 87–93.

Gilbert, A. (2002, April). *How transformative is sociotransformative constructivism? Utilizing critical ethnography to investigate a discourse of transformation.* Paper presented at the annual meeting of the National Association for Research in Science Teaching, New Orleans, LA.

Gilbert, A., & Yerrick, R. (2001). Same school, separate worlds: A sociocultural study of identity, resistance, and negotiation in a rural, lower track science classroom. *Journal of Research in Science Teaching, 38*, 574–598.

Giroux, H. (1992). *Border crossings: Cultural workers and the politics of education.* New York: Routledge.

Goshorn, K. (1996). Social rationality, risk, and the right to know: Information leveraging with the toxic release inventory. *Public Understanding of Science, 5*, 297–320.

Gott, R., Duggan, S., & Johnson, P. (1999). What do practising applied scientists do and what are the implications for science education? *Research in Science & Technological Education, 17*, 97–107.

Grace, M. M., & Ratcliffe, M. (2002). The science and values that young people draw upon to make decisions about biological conservation issues. *International Journal of Science Education, 24*, 1157–1169.

Graube, G., Dyrenfurth, M. J., & Theuerkauf, W. E. (Eds.). (2003). *Technology education: International concepts and perspectives.* New York: Peter Lang.

Gray, B. (1999). Teacher development through curriculum development experiences from the science through applications project. In M. Keogh & S. Naidoo (Eds.), *Proceedings of the 9th symposium of IOSTE*: Vol. 1, pp. 264–272. Durban, South Africa: University of Durban-Westville.

Gray, D., & Bryce, T. (2003, August). *Controversial issues in science education: Implications for the professional development of teachers.* Paper presented at the European Science Education Research Association (ESERA) 2003 conference, Noordwijkerhout, The Netherlands.

Greene, B. (1999). *The elegant universe.* New York: Norton.

Griffard, P., & Wandersee, J. (1999). Challenges to meaningful learning in African-American females at an urban science high school. *International Journal of Science Education, 21*, 611–632.

Haidar, H. H. (2002). Emirates secondary school science teachers' perspectives on the nexus between modern science and Arab culture. *International Journal of Science Education, 24*, 611–626.

Halsey, A. H., Lauder, H., Brown, P., & Wells, A. S. (Eds.). (1997). *Education: Culture, economy, and society.* New York: Oxford University Press.

Hammond, L. (2001). Notes from California: An anthropological approach to urban science education for language minority families. *Journal of Research in Science Teaching, 38*, 983–999.

Hammond, L., & Brandt, C. (2004). Science and cultural process: Defining an anthropological approach to science education. *Studies in Science Education, 40*, 1–47.

Hand, B., & Treagust, D. F. (1997). Monitoring teachers' referents for classroom practice using metaphors. *International Journal of Science Education, 19*, 183–192.

Hansen, K.-H., & Olson, J. (1996). How teachers construe curriculum integration: The science, technology, society (STS) movement as Bildung. *Journal of Curriculum Studies, 28*, 669–682.

Hart, C. (2001). Examining relations of power in a process of curriculum change: The case of VCE physics. *Research in Science Education, 31,* 525–551.

Hart, C. (2002). Framing curriculum discursively: Theoretical perspectives on the experience of VCE physics. *International Journal of Science Education, 24,* 1055–1077.

Hart, C. (2004, July). *The challenge of reconstructing school physics: Reflections from a personal experience as curriculum reformer.* Paper presented to the 11th International Organization for Science & Technology Education (IOSTE) symposium, Lublin, Poland.

Hart, E. P. (1989). Toward renewal of science education: A case study of curriculum policy development. *Science Education, 73,* 607–634.

Hart, E. P., & Robottom, I. M. (1990). The science-technology-movement in science education: A critique of the reform process. *Journal of Research in Science Teaching, 27,* 575–588.

Häussler, P., & Hoffmann, L. (2000). A curricular frame for physics education: Development, comparison with students' interests, and impact on students' achievement and self-concept. *Science Education, 84,* 689–705.

Hawkins, J., & Pea, R. D. (1987). Tools for bridging the cultures of everyday and scientific thinking. *Journal of Research in Science Teaching, 24,* 291–307.

Helms, J. V. (1998). Science—and me: Subject matter and identity in secondary school science teachers. *Journal of Research in Science Teaching, 35,* 811–834.

Hennessy, S. (1993). Situated cognition and cognitive apprenticeship: Implications for classroom learning. *Studies in Science Education, 22,* 1–41.

Hines, J., Hungerford, H., & Tomera, A. (1987). Analysis and synthesis of research on responsible environmental behaviour. *Journal of Environmental Education, 18*(1), 1–8.

Hines, S. M. (2003). *Multicultural science education: Theory, practice, and promise.* New York: Peter Lang.

Hlady, D. L. (1992). *Modulations of an STSE textbook based on teachers' beliefs about the nature of science: A study of teacher practical knowledge.* Unpublished master's thesis, University of Saskatchewan, Saskatoon, Canada.

Hodson, D. (1994). Seeking directions for change: The personalisation and politicisation of science education. *Curriculum Studies, 2,* 71–98.

Hodson, D. (2001). Inclusion without assimilation: Science education from an anthropological and metacognitive perspective. *Canadian Journal of Science, Mathematics and Technology Education, 1,* 161–182.

Hodson, D., Bencze, L., Elshof, L., Pedretti, E., & Nyhof-Young, J. (Eds.). (2002). *Changing science education through action research: Some experiences from the field.* Toronto: University of Toronto Press.

Hofstein, A., Carmi, M., & Ben-Zvi, R. (2003). The development of leadership among chemistry teachers in Israel. *International Journal of Science and Mathematics Education, 1,* 39–64.

Hogan, K. (2000). Exploring a process view of students' knowledge about the nature of science. *Science Education, 84,* 51–70.

Holton, G. (2003). The *Project Physics Course,* then and now. *Science & Education, 12,* 779–786.

Holton, G., Rutherford, J., & Watson, F. (1970). *Project physics course*. New York: Holt, Rinehart & Winston.

Hughes, G. (2000). Marginalization of socioscientific material in science-technology-society science curricula: Some implications for gender inclusivity and curriculum reform. *Journal of Research in Science Teaching, 37*, 426–440.

Hughes, P., More, A. J., & Williams, M. (2004). *Aboriginal ways of learning*. Adelaide, Australia: Flinders Press.

Hunt, A., & Millar, R. (2000). *AS science for public understanding*. Oxford: Heinemann.

Hurd, P. D. (1975). Science, technology and society: New goals for interdisciplinary science teaching. *The Science Teacher, 42*(2), 27–30.

Hurd, P. D. (1986). Perspectives for the reform of science education. *Phi Delta Kappan, 67*, 353–358.

Hurd, P. D. (1989a, February). *A new context for school science education: Science-technology-society*. Paper presented at the Technological Literacy Conference, Washington, DC.

Hurd, P. D. (1989b). Science education and the nation's economy. In A. B. Champagne, B. E. Lovitts, & B. J. Calinger (Eds.), *Scientific literacy* (pp. 15–40). Washington, DC: AAAS.

Hurd, P. D. (1991). Closing the educational gaps between science, technology, and society. *Theory Into Practice, 30*, 251–259.

Hurd, P. D. (1994). Technology and the advancement of knowledge in the sciences. *Bulletin of Science, Technology & Society, 14*, 127–135.

Hurd, P. D. (2002). Modernizing science education. *Journal of Research in Science Teaching, 39*, 3–9.

Inuit Subject Advisory Committee. (1996). *Inuuqatigiit: The curriculum from the Inuit perspective*. Yellowknife, Canada: Northwest Territories Education, Culture and Employment.

Irwin, A. R. (1995). *Citizen science: A study of people, expertise and sustainable development*. New York: Routledge.

Irwin, A. R. (2000). Historical case studies: Teaching the nature of science in context. *Science Education, 84*, 5–26.

Irzik, G. (1998). Philosophy of science and radical intellectual Islam in Turkey. In W. W. Cobern (Ed.), *Socio-cultural perspectives on science education* (pp. 163–179). Boston: Kluwer Academic.

Jeans, S. L. (1998, May). *Teacher images of the intent of science curriculum policy: Experienced and novice teachers at work*. Paper presented at the annual meeting of the Canadian Society for the Study of Education, Ottawa, Ontario, Canada.

Jegede, O. J. (1995). Collateral learning and the eco-cultural paradigm in science and mathematics education in Africa. *Studies in Science Education, 25*, 97–137.

Jegede, O. J., & Aikenhead, G. S. (1999). Transcending cultural borders: Implications for science teaching. *Research in Science and Technology Education, 17*, 45–66.

Jegede, O. J., & Okebukola, P. A. (1991). The relationship between African traditional cosmology and students' acquisition of a science process skill. *International Journal of Science Education, 13*, 37–47.

Jenkins, E. (1985). History of science education. In T. Husén & T. N. Postlethwaite (Eds.), *International encyclopedia of education* (pp. 4453–4456). Oxford: Pergamon Press.

Jenkins, E. (1992). School science education: Towards a reconstruction. *Journal of Curriculum Studies, 24,* 229–246.

Jenkins, E. (1999). School science, citizenship and the public understanding of science. *International Journal of Science Education, 21,* 703–710.

Jenkins, E. (2000). "Science for all": Time for a paradigm shift? In R. Millar, J. Leach, & J. Osborne (Eds.), *Improving science education: The contribution of research* (pp. 207–226). Buckingham, UK: Open University Press.

Jenkins, E. (2004). Science education: Research, practice and policy. In E. Scanlon, P. Murphy, J. Thomas, & E. Whitelegg (Eds.), *Reconsidering science learning* (pp. 235–249). New York: RoutledgeFalmer.

Jiménez-Aleizandre, M-P., Agraso, M. F., & Eirexas, F. (2004, April). *Scientific authority and empirical data in argument warrants and the Prestige oil spill.* Paper presented at the annual meeting of the National Association for Research in Science Teaching, Vancouver, British Columbia, Canada.

Jiménez-Aleizandre, M-P., & Pereiro-Muñoz, C. (2002). Knowledge producers or knowledge consumers? Argumentation and decision making about environmental management. *International Journal of Science Education, 24,* 1171–1190.

Kawagley, A. O. (1990). Yup'ik ways of knowing. *Canadian Journal of Native Education, 17*(2), 5–17.

Kawagley, A. O. (1995). *A Yupiaq worldview.* Prospect Heights, IL: Waveland Press.

Kawagley, A. O., Norris-Tull, D., & Norris-Tull, R. A. (1998). The indigenous worldview of Yupiaq culture: Its scientific nature and relevance to the practice and teaching of science. *Journal of Research in Science Teaching, 35,* 133–144.

Kawasaki, K. (1996). The concepts of science in Japanese and Western education. *Science & Education, 5,* 1–20.

Kawasaki, K. (2002). A cross-cultural comparison of English and Japanese linguistic assumptions influencing pupils' learning of science. *Canadian and International Education, 31,* 19–51.

Keane, M., & Malcolm, C. (2003). Relevant science education, but relevant to what? *LabTalk, 47*(2), 4–9. Victoria, Australia: Science Teachers' Association of Victoria.

Kearney, M. (1984). *World view.* Novato, CA: Chandler & Sharp.

Keiny, S. (1993). School-based curriculum development as a process of teachers' professional development. *Educational Action Research, 1,* 65–93.

Keiny, S. (1996). A community of learners: Promoting teachers to become learners. *Teachers and Teaching: Theory and Practice, 2,* 243–272.

Keiny, S. (1999). A reflective inquiry of STS learning-as-inquiry. In M. Keogh & S. Naidoo (Eds.), *Proceedings of the 9th symposium of IOSTE*: Vol. 1, p. 347. Durban, South Africa: University of Durban-Westville.

Kelly, G. J., Carlsen, W. S., & Cunningham, C. M. (1993). Science education in sociocultural context: Perspectives from the sociology of science. *Science Education, 77,* 207–220.

Kempton, W., Boster, J. S., & Hartley, J. A. (1995). *Environmental values in American culture*. Cambridge, MA: MIT Press.

Khishfe, R., & Abd-El-Khalick, F. (2002). Influence of explicit and reflective versus implicit inquiry-oriented instruction on sixth graders' views of nature of science. *Journal of Research in Science Teaching, 39*, 551–578.

Kilian-Schrum, S. (1996). An investigation into the implementation of a high school science program with an STS dimension. In K. Calhoun, R. Panwar, & S. Shrum (Eds.), *Proceedings of the 8th symposium of IOSTE*: Vol. 1, pp. 104–108. Edmonton, Canada: University of Alberta, Faculty of Education.

Kincheloe, J. L. (1998). Critical research in science education. In B. J. Fraser & K. G. Tobin (Eds.), *International handbook of science education* (pp. 1191–1205). Dordrecht, The Netherlands: Kluwer Academic.

Kleine, K. (1997, June). *The role of past training as a scientist on four middle school science teachers*. Paper presented at the History & Philosophy of Science and Science Teaching Conference, Calgary, Alberta, Canada.

Kliebard, H. M. (1979). The drive for curriculum change in the United States, 1890–1958. I—The ideological roots of curriculum as a field of specialization. *Journal of Curriculum Studies, 11*, 191–202.

Klopfer, L. E. (1969). The teaching of science and the history of science. *Journal of Research in Science Teaching, 6*, 87–95.

Klopfer, L. E. (1992). An historical perspective on the history and nature of science in school science programs. In R. Bybee, J. D. Ellis, U. R. Giese, & L. Parsi (Eds.), *Teaching about the history and nature of science and technology: Background papers* (pp. 105–130). Colorado Springs, CO: BSCS.

Klopfer, L. E., & Champagne, A. B. (1990). Ghosts of crisis past. *Science Education, 74*, 133–154.

Klopfer, L. E., & Cooley, W. W. (1963). "The History of Science Cases" for high school in the development of student understanding of science and scientists. *Journal of Research in Science Teaching, 1*, 33–47.

Klopfer, L. E., & Watson, F. G. (1957). Historical materials and high school science teaching. *The Science Teacher, 24*, 264–293.

Knain, E. (2001). Ideologies in school science textbooks. *International Journal of Science Education, 23*, 319–329.

Knain, E. (2005). Identity and genre literacy in high-school students' experimental reports. *International Journal of Science Education, 27*, 607–624.

Knudtson, P., & Suzuki, D. (1992). *Wisdom of the elders*. Toronto, Canada: Stoddart.

Kolstø, S. D. (2000). Consensus projects: Teaching science for citizenship. *International Journal of Science Education, 22*, 645–664.

Kolstø, S. D. (2001a). *Science education for citizenship: Thoughtful decision-making about science-related social issues*. Unpublished doctoral dissertation, University of Oslo, Faculty of Mathematics and Natural Sciences, Oslo, Norway.

Kolstø, S. D. (2001b). "To trust or not to trust, . . ."—Pupils' ways of judging information encountered in a socio-scientific issue. *International Journal of Science Education, 23*, 877–901.

Kolstø, S. D., Kristensen, T., Arnesen, E., Mathiassen, K., Mestad, I., Quale, A., et al. (2004, April). *Science students' critical examination of scientific information related to socioscientific issues.* Paper presented at the annual meeting of the National Association for Research in Science Teaching, Vancouver, British Columbia, Canada.

Koosimile, A. T. (2004). Out-of-school experiences in science classes: Problems, issues and challenges in Botswana. *International Journal of Science Education, 26,* 483–496.

Kortland, J. (1992). Environmental education: Sustainable development and decision-making. In R. E. Yager (Ed.), *The status of science-technology-society reform efforts around the world.* ICASE 1992 Yearbook (pp. 32–39). Knapp Hill, South Harting, Petersfield, England: International Council of Associations for Science Education.

Kortland, J. (1996). An STS case study about students' decision making on the waste issue. *Science Education, 80,* 673–689.

Kortland, J. (2001). *A problem posing approach to teaching decision making about the waste issue.* Utrecht, The Netherlands: University of Utrecht Cdβ Press.

Koul, R. (2003). Revivalist thinking and student conceptualizations of science/religion. *Studies in Science Education, 39,* 103–124.

Kozoll, R. H., & Osborne, M. D. (2004). Finding meaning in science: Lifeworld, identity, and self. *Science Education, 88,* 157–181.

Krasilchik, M. (1995). The ecology of science education: Brazil: 1950–90. *International Journal of Science Education, 17,* 413–423.

Krogh, L. B., & Thomsen, P. V. (2005). Studying students' attitudes towards science from a cultural perspective but with a quantitative methodology: Border crossing into the physics classroom. *International Journal of Science Education, 27,* 281–302.

Krugly-Smolska, E. (1994). An examination of some difficulties in integrating Western science into societies with an indigenous scientific tradition. *Interchange, 25,* 325–334.

Krugly-Smolska, E. (1995). Cultural influences in science education. *International Journal of Science Education, 17,* 45–58.

Krull-Romanyshyn, B. (1996). Teachers and curriculum change: Examining an inservice program implementation. In K. Calhoun, R. Panwar, & S. Shrum (Eds.), *Proceedings of the 8th symposium of IOSTE*: Vol. 1, pp. 118–127. Edmonton, Canada: University of Alberta, Faculty of Education.

Kuhn, T. (1962). *The structure of scientific revolutions.* Chicago: University of Chicago Press.

Kumar, D. D., & Chubin, D. E. (Eds.). (2000). *Science, technology, and society: A sourcebook on research and practice.* New York: Kluwer Academic/Plenum.

Kyle, W. C. (1995). Editorial: Science, science education, and the public. *Journal of Research in Science Teaching, 32,* 551–553.

Kyle, W. C. (1996). Editorial: The importance of investing in human resources. *Journal of Research in Science Teaching, 33,* 1–4.

Kyle, W. C. (Director). (2004). *Science, technology & culture.* Retrieved December 1, 2004, from http://stc.umsl.edu.

Lantz, O., & Kass, H. (1987). Chemistry teachers' functional paradigms. *Science Education, 71,* 117–134.

Larochelle, M., & Désautels, J. (1991). "Of course, it's just obvious": Adolescents' ideas of scientific knowledge. *International Journal of Science Education, 13,* 373–389.

Larson, J. O. (1995, April). *Fatima's rules and other elements of an unintended chemistry curriculum.* Paper presented to the American Educational Research Association Annual Meeting, San Francisco, CA.

Law, N. (2002). Scientific literacy: Charting the terrains of a multifaceted enterprise. *Canadian Journal of Science, Mathematics and Technology Education, 2,* 151–176.

Law, N., Fensham, P. J., Li, S., & Wei, B. (2000). Public understanding of science as basic literacy. In R. T. Cross & P. J. Fensham (Eds.), *Science and the citizen for educators and the public* (pp. 145–155). Melbourne: Arena.

Lawrence, N., & Eisenhart, M. (2002). The language of science and the meaning of abortion. In W-M. Roth & J. Désautels (Eds.), *Science education as/for sociopolitical action* (pp. 185–206). New York: Peter Lang.

Lawrenz, F., & Gray, B. (1995). Investigation of worldview theory in a South African context. *Journal of Research in Science Teaching, 32,* 555–568.

Layton, D. (1973). *Science for the people: The origins of the school science curriculum in England.* London: George Allen & Unwin.

Layton, D. (1981). The schooling of science in England, 1854–1939. In R. MacLeod & P. Collins (Eds.), *The parliament of science* (pp. 188–210). Northwood, England: Science Reviews.

Layton, D. (1986). Science education and values education—An essential tension. In J. Brown, A. Cooper, T. Horton, F. Toates, & D. Zeldin (Eds.), *Science in schools* (pp. 110–120). Milton Keynes, UK: Open University Press.

Layton, D. (1991). Science education and praxis: The relationship of school science to practical action. *Studies in Science Education, 19,* 43–79.

Layton, D., Jenkins, E., Macgill, S., & Davey, A. (1993). *Inarticulate science? Perspectives on the public understanding of science and some implications for science education.* Driffield, East Yorkshire, England: Studies in Education.

Leach, J., Driver, R., Millar, R., & Scott, P. (1997). A study of progression in learning about "the nature of science": Issues of conceptualization and methodology. *International Journal of Science Education, 19,* 147–166.

Leblanc, R. (1989). *Department of education summer science institute.* Halifax, Nova Scotia, Canada: Ministry of Education, PO Box 578.

Lederman, N. G. (1992). Students' and teachers' conceptions of the nature of science: A review of the research. *Journal of Research in Science Teaching, 29,* 331–359.

Lederman, N. G., Abd-El-Khalick, F., Bell, R. L., & Schwartz, R. S. (2002). Views of nature of science questionnaire: Toward valid and meaningful assessment of learners' conceptions of nature of science. *Journal of Research in Science Teaching, 39,* 497–521.

Lee, H., Choi, K., & Abd-El-Khalick, F. (2003, March). *Korean science teachers' perceptions of the introduction of socioscientific issues into the science curriculum.* Paper presented at the annual meeting of the National Association for Research in Science Teaching, Philadelphia, PA.

Lee, J. D. (2002). More than ability: Gender and personal relationships influence science and technology involvement. *Sociology of Education, 74*, 349–374.

Lee, O. (1997). Diversity and equity for Asian American students in science education. *Science Education, 81*, 107–122.

Lee, O. (2002). Promoting scientific inquiry with elementary students from diverse cultures and languages. *Review of Research in Education, 26*, 23–69.

Lee, O., & Fradd, S. H. (1998). Science for all, including students from non-English-language backgrounds. *Educational Researcher, 27*(4), 12–21.

Lee, S., & Roth, W-M. (2002). Learning science in the community. In W-M. Roth & J. Désautels (Eds.), *Science education as/for sociopolitical action* (pp. 37–66). New York: Peter Lang.

Lemke, J. L. (2001). Articulating communities: Sociocultural perspectives on science education. *Journal of Research in Science Teaching, 38*, 296–316.

Lemmer, M., Lemmer, T. N., & Smit, J. J. A. (2003). South African students' views of the universe. *International Journal of Science Education, 25*, 563–582.

Levinson, R. (2004). Teaching bioethics in science: Crossing a bridge too far? *Canadian Journal of Science, Mathematics and Technology Education, 4*, 353–369.

Levinson, R., & Turner, S. (2001). *Valuable lessons: Engaging with the social context of science in schools.* London: The Wellcome Trust.

Lijnse, P. (1990). Energy between the life-world of pupils and the world of physics. *Science Education, 74*, 571–583.

Lijnse, P. (1995). "Developmental research" as a way to an empirically based "didactical structure" of science. *Science Education, 79*, 189–199.

Lijnse, P., Kortland, K., Eijkelhof, H., van Genderen, D., & Hooymayers, H. (1990). A thematic physics curriculum: A balance between contradictory curriculum forces. *Science Education, 74*, 95–103.

Lin, H-S. (1998, April). *Promoting pre-service science teachers' understanding about the nature of science through history of science.* Paper presented at the annual meeting of the National Association for Research in Science Teaching, San Diego, CA.

Lin, H-S., Hung, J-Y., & Hung, S-C. (2002). Using the history of science to promote students' problem-solving ability. *International Journal of Science Education, 24*, 453–464.

Lindahl, B. (2003). *Lust att lära naturvetenskap och teknik? En longitudinell studie om vägen till gymnasiet.* [Pupils' responses to school science and technology? A longitudinal study of pathways to upper secondary school.]. Unpublished doctoral dissertation (in Swedish with an English summary), Göteborg Studies in Educational Sciences 196, Acta Universitatis Gothoburgensis, Göteborg, Sweden.

Linn, M. C., Clark, D., & Slotta, J. D. (2003). WISE design for knowledge integration. *Science Education, 87*, 781–796.

Loo, S. P. (2001). Islam, science and science education: Conflict or concord? *Studies in Science Education, 36*, 45–78.

Lottero-Perdue, P. S., & Brickhouse, N. W. (2002). Learning on the job: The acquisition of scientific competence. *Science Education, 86*, 756–782.

Louden, W., & Wallace, J. (1994). Knowing and teaching science: The constructivist paradox. *International Journal of Science Education, 16,* 649–657.

Loughran, J., & Derry, N. (1997). Researching teaching for understanding: The students' perspective. *International Journal of Science Education, 19,* 925–938.

Lubben, F., & Campbell, B. (1996). Contextualizing science teaching in Swaziland: Some student reactions. *International Journal of Science Education, 18,* 311–320.

Luft, J. A. (1999). The border crossings of a multicultural science education enthusiast. *School Science and Mathematics, 99,* 380–388.

Luft, J. A. (2001). Changing inquiry practices and beliefs: The impact of an inquiry-based professional development programme on beginning and experienced secondary science teachers. *International Journal of Science Education, 23,* 517–534.

Lugones, M. (1987). Playfulness, "world"-travelling, and loving perception. *Hypatia, 2*(2), 3–19.

Lumpe, T., Haney, J. J., & Czerniak, C. M. (1998). Science teacher beliefs and intentions to implement science-technology-society (STS) in the classroom. *Journal of Science Teacher Education, 9,* 1–24.

Lynch, P. P. (1996). Students' alternative frameworks for the nature of matter: A cross-cultural study of linguistic and cultural interpretations. *International Journal of Science Education, 18,* 743–752.

Lynch, S. (2001). "Science for all" is not equal to "one size fits all": Linguistic and cultural diversity and science education reform. *Journal of Research in Science Teaching, 38,* 622–627.

Lyons, T. S. (2003). *Decisions by "science proficient" year 10 students about post-compulsory high school science enrolment: A sociocultural exploration.* Unpublished doctoral dissertation, University of New England, Armidale, NSW, Australia.

MacIvor, M. (1995). Redefining science education for Aboriginal students. In M. Battiste & J. Barman (Eds.), *First Nations education in Canada: The circle unfolds* (pp. 73–98). Vancouver, Canada: University of British Columbia Press.

MacLeod, R. (1981). Introduction: On the advancement of science. In R. MacLeod & P. Collins (Eds.), *The parliament of science* (pp. 17–42). Northwood, England: Science Reviews.

Maddock, M. N. (1981). Science education: An anthropological viewpoint. *Studies in Science Education, 8,* 1–26.

Malcolm, C. (2004). Thoughts from South Africa: Context, relevance and interest. *LabTalk, 48*(1), 31–34. Victoria, Australia: Science Teachers' Association of Victoria.

Malcolm, C., & Alant, B. (2004). Finding direction when the ground is moving: Science education research in South Africa. *Studies in Science Education, 40,* 49–104.

Malcolm, C., & Keogh, M. (2004). The science teacher as curriculum developer: Do you think it will rain today? In B. Gray, P. Naidoo, & M. Savage (Eds.), *School science in Africa: Teaching to learn, learning to teach* (pp. 105–128). Cape Town, South Africa: AFCLIST/Juta.

Manassero-Mas, M. A., & Vázquez-Alonso, A. (1998). *Opinions sobre ciencia, tecnologia i societat*. Les Illes Balears, Spain: Conselleria d'Educació i Cultura i Esports.

Manassero-Mas, M. A., Vázquez-Alonso, A., & Acevedo-Díaz, J. A. (2001). *Avaluació dels temes de ciencia, tecnologia i societat*. Les Illes Balears, Spain: Conselleria d'Educació i Cultura del Govern de les Illes Balears.

Marchetti, D., & Camino, E. (2003, August). *Experimentation of interactive didactic activities on complex and controversial environmental issues with students of higher secondary school*. Paper presented at the European Science Education Research Association (ESERA) 2003 conference, Noordwijkerhout, The Netherlands.

Mason, D. (1996, March/April). *Life after ChemCom: Do they succeed in university-level chemistry courses?* Paper presented at the annual meeting of the National Association for Research in Science Teaching, St. Louis, MO. (ERIC Document Reproduction Service No. ED393693)

Matthews, M. R. (1994). *Science teaching: The role of history and philosophy of science*. New York: Routledge.

May, S. (Ed.). (1999). *Indigenous community-based education*. Philadelphia: Multilingual Matters.

Mayoh, K., & Knutton, S. (1997). Using out-of-school experience in science lessons: Reality or rhetoric? *International Journal of Science Education, 19*, 849–867.

Mbajiorgu, N. M., & Ali, A. (2003). Relationship between STS approach, scientific literacy, and achievement in biology. *Science Education, 87*, 31–39.

McClelland, L. W. (1998, May). *Curriculum change: What experienced science teachers say about it*. Paper presented at the annual meeting of the Canadian Society for the Study of Education, Ottawa, Ontario, Canada.

McComas, W. F. (Ed.). (1998). *The nature of science in science education: Rationales and strategies*. Dordrecht, The Netherlands: Kluwer Academic.

McComas, W. F., Clough, M. P., & Almazora, H. (1998). The nature of science in science education: An introduction. *Science & Education, 7*, 511–532.

McConnell, M. C. (1982). Teaching about science, technology and society at the secondary school level in the United States: An education dilemma for the 1980s. *Studies in Science Education, 9*, 1–32.

McFadden, C. (1991). Toward an STS school curriculum. *Science Education, 75*, 457–469.

McFadden, C. (1996). The Atlantic science curriculum project. In K. Calhoun, R. Panwar, & S. Shrum (Eds.), *Proceedings of the 8th symposium of IOSTE*: Vol. 3, pp. 156–161. Edmonton, Canada: University of Alberta, Faculty of Education.

McFadden, C., & Yager, R. E. (1997). *SciencePlus technology and society*. Austin, TX: Holt, Rinehart & Winston.

McGinn, M. K., & Roth, W-M. (1999). Preparing students for competent scientific practice: Implication of recent research in science and technology studies. *Educational Researcher, 28*(3), 14–24.

McGinnis, J. R., & Simmons, P. (1999). Teachers' perspectives of teaching science-technology-society in local cultures: A socio-cultural analysis. *Science Education, 83*, 179–211.

McGrath, C., & Watts, M. (1996). SATIS factions: Approaches to relevance in science and technology education. In K. Calhoun, R. Panwar, & S. Shrum (Eds.), *Proceedings of the 8th symposium of IOSTE*: Vol. 3, pp. 162–167. Edmonton, Canada: University of Alberta, Faculty of Education.

McKinley, E. (1996). Towards an indigenous science curriculum. *Research in Science Education, 26*, 155–167.

McKinley, E. (2001). Cultural diversity: Masking power with innocence. *Science Education, 85*, 74–76.

McKinley, E. (2003). *Brown bodies, white coats: Postcolonialism, Māori women, and science*. Unpublished doctoral dissertation, University of Waikato, Hamilton, Aotearoa New Zealand.

McKinley, E. (2005). Locating the global: Culture, language and science education for indigenous students. *International Journal of Science Education, 27*, 227–241.

McKinley, E., McPherson Waiti, P., & Bell, B. (1992). Language, culture and science education. *International Journal of Science Education, 14*, 579–595.

McKinley, E., Stewart, G., & Richards, P. (2004). *Māori knowledge, language and participation in mathematics and science education*. (Final Report). Hamilton, Aotearoa New Zealand: University of Waikato, School of Education.

McRobbie, C., & Tobin, K. (1995). Restraints to reform: The congruence of teacher and student actions in a chemistry classroom. *Journal of Research in Science Teaching, 32*, 373–385.

McSharry, G., & Jones, S. (2002). Television programming and advertisements: Help or hindrance to effective science education? *International Journal of Science Education, 24*, 487–497.

Medicine, B. (2000). Afterward. In M. Benham & J. Cooper (Eds.), *Indigenous educational models for contemporary practice: In our mother's voice* (pp. 159–161). Mahwah, NJ: L. Erlbaum Associates.

Medvitz, A. G. (1996). Science, schools and culture: The complexity of reform in science education. In K. Calhoun, R. Panwar, & S. Shrum (Eds.), *Proceedings of the 8th symposium of IOSTE*: Vol. 2, pp. 158–163. Edmonton, Canada: University of Alberta, Faculty of Education.

Mendelsohn, E. (1976). Values and science: A critical reassessment. *The Science Teacher, 43*(1), 20–23.

Mendelsohn, E., & Elkana, Y. (Eds.). (1981). *Sciences and cultures: Anthropological and historical studies of the sciences* (Sociology of the Sciences Yearbook, Vol. 5). Boston: Reidel.

Menzies, C. R., Archibald, Q. Q. J., & Smith, G. H. (2004). Editorial: Transformational sites of indigenous education. *Canadian Journal of Native Education, 28*, 1–7.

Meyer, K. (1998). Reflections on being female in school science: Toward a praxis of teaching science. *Journal of Research in Science Teaching, 35*, 463–471.

Michael, M. (1992). Lay discourses of science, science-in-general, science-in-particular and self. *Science, Technology & Human Values, 17*, 313–333.

Michie, M. (2004, April). *Teaching science to Indigenous students: Teacher as culture broker or is it something else?* Paper presented at the annual meeting of the

National Association for Research in Science Teaching, Vancouver, British Columbia, Canada.

Millar, R. (2000). Science for public understanding: Developing a new course for 16–18 year old students. In R. T. Cross & P. J. Fensham (Eds.), *Science and the citizen for educators and the public* (pp. 201–214). Melbourne: Arena.

Millar, R., Leach, J., & Osborne J. (Eds.). (2000). *Improving science education: The contribution of research.* Birmingham, UK: Open University Press.

Millar, R., & Osborne, J. (Eds.). (1998). *Beyond 2000: Science education for the future.* London: King's College, School of Education.

Milne, C. E. (1998). Philosophically correct science stories? Examining the implications of heroic science stories for school science. *Journal of Research in Science Teaching, 35,* 175–187.

Milne, C. E., & Taylor, P. C. (1998). Between myth and a hard place. In W. W. Cobern (Ed.), *Socio-cultural perspectives on science education* (pp. 25–48). Boston: Kluwer Academic.

Mitchener, C. P., & Anderson, R. D. (1989). Teachers' perspective: Developing and implementing an STS curriculum. *Journal of Research in Science Teaching, 26,* 351–369.

Moje, E. B., Callazo, T., Carrillo, R., & Marx, R. W. (2001). "Maestro, what is 'quality'?": Language, literacy, and discourse in project-based science. *Journal of Research in Science Teaching, 38,* 469–498.

Monk, M., & Osborne, J. (1997). Placing the history and philosophy of science on the curriculum: A model for the development of pedagogy. *Science Education, 81,* 405–424.

Moss, D. M., Abrams, E. D., & Robb, J. (2001). Examining student conceptions of the nature of science. *International Journal of Science Education, 23,* 771–790.

Mulholland, J., & Wallace, J. (2003). Crossing borders: Learning and teaching primary science in the preservice to inservice transition. *International Journal of Science Education, 25,* 879–898.

Munby, H., Cunningham, M., & Lock, C. (2000). School science culture: A case study of barriers to developing professional knowledge. *Science Education, 84,* 193–211.

National Research Council [NRC]. (1996). *National science education standards.* Washington, DC: National Academy Press.

Nelkin, D. (1995). *Selling science: How the press covers science and technology* (Rev. ed.). New York: Freeman.

Nelson-Barber, S., & Estrin, E. T. (1995). *Culturally responsive mathematics and science education for Native students.* San Francisco: Far West Laboratory for Education Research and Development.

Nicol, C., Tsai, L-L., & Gaskell, J. (2004). Students and applied academics: Learner agency in a changing curriculum. *Canadian Journal of Science, Mathematics and Technology Education, 4,* 209–221.

Nieswandt, M., & Bellomo, K. (2003, March). *Persistence and change in secondary science preservice teacher's personal belief system about science teaching.* Paper presented at the annual meeting of the National Association for Research in Science Teaching, Philadelphia, PA.

Ninnes, P. (1995). Informal learning contexts in Solomon Islands and their implications for the cross-cultural classroom. *International Journal of Educational Development, 15,* 15–26.

Ninnes, P. (2000). Representations of indigenous knowledges in secondary school science textbooks in Australia and Canada. *International Journal of Science Education, 22,* 603–617.

Ninnes, P. (2004). Discourses of cultural diversity in the science curriculum: Connections, contradictions, and colonialisms. *Discourse: Studies in the Cultural Politics of Education, 25,* 261–278.

Norman, O., Ault, C. R., Bentz, B., & Meskimen, L. (2001). The Black–White "achievement gap" as a perennial challenge of urban science education: A sociocultural and historical overview with implication for research and practice. *Journal of Research in Science Teaching, 38,* 1101–1114.

Novak, J. D. (2002). Meaningful learning: The essential factor for conceptual change in limited or appropriate propositional hierarchies (LIPHs) leading to empowerment of learners. *Science Education, 86,* 548–571.

Novak, J. D. (2004). Reflections on a half-century of thinking in science education and research: Implications from a twelve-year longitudinal study of children's learning. *Canadian Journal of Science, Mathematics and Technology Education, 4,* 23–41.

Ødegaard, M. (2003a). Dramatic science: A crucial review of drama in science education. *Studies in Science Education, 39,* 75–101.

Ødegaard, M. (2003b, August). *"What will our baby say?" Biotechnology, role-play and the ethics of caring.* Paper presented at the European Science Education Research Association (ESERA) 2003 conference, Noordwijkerhout, The Netherlands.

Office of Educational Research and Improvement. (1986). *National assessment of educational progress: Science report card.* Washington, DC: Author.

Ogawa, M. (1995). Science education in a multi-science perspective. *Science Education, 79,* 583–593.

Ogawa, M. (1997). Toward an epic description of science education: Cultural history of science education in Japan. In M. Ogawa (Ed.), *Effects of traditional cosmology on science education* (pp. 96–125). Ibaraki, Japan: Ibaraki University, Faculty of Education.

Ogawa, M. (2004a, April). *Origin, structure and nature of indigenous science and "seigyo" (subsistence).* Paper presented at the annual meeting of the National Association for Research in Science Teaching, Vancouver, British Columbia, Canada.

Ogawa, M. (2004b, July). *How did Japanese students respond to the questionnaire of "the relevance of science education"?* Paper presented to the 11th International Organization for Science & Technology Education (IOSTE) symposium, Lublin, Poland.

Ogborn, J. (2002). Ownership and transformation: Teachers using curriculum innovations. *Physics Education, 37,* 142–146.

Ogunniyi, M. B. (1988). Adapting western science to traditional African culture. *International Journal of Science Education, 10,* 1–9.

Ogunniyi, M. B. (1996). Science, technology and mathematics: The problem of developing critical human capital in Africa. *International Journal of Science Education, 18,* 267–284.

Ogunniyi, M. B., Jegede, O. J., Ogawa, M., Yandila, C. D., & Oladele, F. K. (1995). Nature of worldview presuppositions among science teachers in Botswana, Indonesia, Japan, Nigeria, and the Philippines. *Journal of Research in Science Teaching, 32*, 817–831.

Okebukola, P. A. O., & Jegede, O. J. (1990). Eco-cultural influences upon students' concept attainment in science. *Journal of Research in Science Teaching, 27*, 661–669.

O'Loughlin, M. (1992). Rethinking science education: Beyond Piagetian constructivism toward a sociocultural model of teaching and learning. *Journal of Research in Science Teaching, 29*, 791–820.

Orange, A. D. (1981). The beginnings of the British Association, 1831–1851. In R. MacLeod & P. Collins (Eds.), *The parliament of science* (pp. 43–64). Northwood, England: Science Reviews.

Orpwood, G. (1985). Toward the renewal of Canadian science education. I. Deliberative inquiry model. *Science Education, 69*, 477–489.

Osborne, J., & Collins, S. (2000). *Pupils' and parents' views of the school science curriculum.* London: Kings College London.

Osborne, J., & Collins, S. (2001). Pupils' views of the role and value of the science curriculum: A focus group study. *International Journal of Science Education, 23*, 441–467.

Osborne, J., Collins, S., Ratcliffe, M., Millar, R., & Duschl, R. (2003). What "ideas-about-science" should be taught in school science? A Delphi study of the expert community. *Journal of Research in Science Teaching, 40*, 692–720.

Osborne, J., Duschl, R., & Fairbrother, B. (2003, March). *Breaking the mould? Teaching science for public understanding: Lessons from the classroom.* Paper presented at the annual meeting of the National Association for Research in Science Teaching, Philadelphia, PA.

Östman, L. (1996). Discourses, discursive meanings and socialization in chemical education. *Journal of Curriculum Studies, 28*, 293–304.

O'Sullivan, E. V., Morrell, A., & O'Connor, M. S. (Eds.). (2002). *Expanding the boundaries of transformative learning: Essays on theory and praxis.* New York: Palgrave.

Oulton, C., Dillon, J., & Grace, M. M. (2004). Reconceptualizing the teaching of controversial issues. *International Journal of Science Education, 26*, 411–423.

Oxford University Department of Educational Studies. (1989). *Enquiry into the attitudes of sixth-formers towards choice of science and technology courses in higher education.* Oxford, England: Author.

Panwar, R., & Hoddinott, J. (1995). The influence of academic scientists and technologists on Alberta's science curriculum policy and programme. *International Journal of Science Education, 17*, 505–518.

Patronis, T., Potari, D., & Spiliotopoulou, V. (1999). Students' argumentation in decision-making on a socio-scientific issue: Implications for teaching. *International Journal of Science Education, 21*, 745–754.

Pedersen, J. E., & Totten, S. (2001). Beliefs of science teachers toward the teaching of science/technological/social issues: Are we addressing national standards? *Bulletin of Science, Technology & Society, 21*, 376–393.

Pedretti, E. (1999). Decision making and STS education: Exploring scientific knowledge and social responsibility in schools and science centers through an issues-based approach. *School Science and Mathematics, 99*, 174–181.

Pedretti, E. (2003). Teaching science, technology, society and environment (STSE) education. In D. L. Zeidler (Ed.), *The role of moral reasoning on socioscientific issues and discourse in science education* (pp. 219–239). Boston: Kluwer Academic.

Pedretti, E. (2004). Perspectives on learning through research on critical issues-based science center exhibitions. *Science Education, 88*(Suppl. 1), S34–S47.

Pedretti, E., Bencze, L., Hewitt, J., Jawani, A., & Romkey, L. (2004, April). *Protocols and paradoxes in STSE education: Using multimedia cases in science teacher preparation.* Paper presented at the annual meeting of the National Association for Research in Science Teaching, Vancouver, British Columbia, Canada.

Pedretti, E., & Hodson, D. (1995). From rhetoric to action: Implementing STS education through action research. *Journal of Research in Science Teaching, 32*, 463–485.

Peshkin, A. (2000). The nature of interpretation in qualitative research. *Educational Researcher, 29*(9), 5–9.

Phelan, P., Davidson, A., & Cao, H. (1991). Students' multiple worlds: Negotiating the boundaries of family, peer, and school cultures. *Anthropology and Education Quarterly, 22*, 224–250.

Pickering, A. (Ed.). (1992). *Science as practice and culture.* Chicago: University of Chicago Press.

Pillay, A. (1996). Case studies of science and technology education programmes in South Africa designed for community problem solving in the environment. In K. Calhoun, R. Panwar, & S. Shrum (Eds.), *Proceedings of the 8th symposium of IOSTE*: Vol. 3, pp. 218–222. Edmonton, Canada: University of Alberta, Faculty of Education.

Pomeroy, D. (1994). Science education and cultural diversity: Mapping the field. *Studies in Science Education, 24*, 49–73.

Posch, P. (1993). The environmental and schools initiatives (ENSI) I: Action research in environmental education. *Educational Action Research, 1*, 447–486.

Prelle, S., & Solomon, J. (1996). Young people's "General Approach" to environmental issues in England and Germany. *Compare, 26*, 91–103.

Quinn, D. (1992). *Ishmael.* New York: Bantam/Turner Books.

Rafea, A. M. (1999). *Power, curriculum making and actor-network theory: The case of physics, technology and society curriculum in Bahrain.* Unpublished doctoral dissertation, University of British Columbia, Vancouver, Canada.

Rahm, J. (2002). Emergent learning opportunities in an inner-city youth gardening program. *Journal of Research in Science Teaching, 39*, 164–184.

Ramsden, J. M. (1992). If it's enjoyable, is it science? *School Science Review, 73*(265), 65–71.

Ramsden, J. M. (1997). How does a context-based approach influence understanding of key chemical ideas at 16+? *International Journal of Science Education, 19*, 697–710.

Rashed, R. (1997). Science as a western phenomenon. In H. Selin (Ed.), *Encyclopaedia of the history of science, technology, and medicine in non-Western cultures* (pp. 84–890). Boston: Kluwer Academic.

Ratcliffe, M. (1997a, September). *Assessing pupils' abilities in dealing with ethical and social aspects of curriculum.* Paper presented at the biennial meeting of the European Science Education Research Association, Rome, Italy.

Ratcliffe, M. (1997b). Pupil decision-making about socio-scientific issues within the science curriculum. *International Journal of Science Education, 19,* 167–182.

Ratcliffe, M. (1999). Evaluation of abilities in interpreting media reports of scientific research. *International Journal of Science Education, 21,* 1085–1099.

Ratcliffe, M. (2004, April). *Science and sustainability: Evaluation of integration of science concepts with global issues.* Paper presented at the annual meeting of the National Association for Research in Science Teaching, Vancouver, British Columbia, Canada.

Ratcliffe, M., Bartholomew, H., Hames, V., Hind, A., Leach, J., Millar, R., et al. (2003, March). *Evidence-based practice in science education: The research-user interface.* Paper presented at the annual meeting of the National Association for Research in Science Teaching, Philadelphia, PA.

Ratcliffe, M., & Grace, M. (2003). *Science education for citizenship: Teaching socio-scientific issues.* Philadelphia: Open University Press.

Ravetz, J. (1997). Simple scientific truths and uncertain policy realities: Implications for science education. *Studies in Science Education, 30,* 5–18.

Read, T. (1998). Kormilda science project. In M. Michie (Ed.), *Science education: Beyond the horizon* (pp. 155–157). (Conasta conference proceedings.) Darwin, Australia: Northern Territory University.

Read, T. (2002). *The Kormilda science project: An indigenous perspective on the earth sciences.* Mount Eliza, Australia: Author.

Reis, P., & Galvão, C. (2004). The impact of socio-scientific controversies in Portuguese natural science teachers' conceptions and practices. *Research in Science Education, 34,* 153–171.

Reiss, M. J. (1999). Teaching ethics in science. *Studies in Science Education, 34,* 115–140.

Reiss, M. J. (2000). *Understanding science lessons: Five years of science teaching.* Milton Keynes, UK: Open University Press.

Reiss, M. J. (2004). Students' attitudes towards science: A long-term perspective. *Canadian Journal of Science, Mathematics and Technology Education, 4,* 97–109.

Rennie, L. J., & Johnston, D. J. (2004). The nature of learning and its implications for research on learning from museums. *Science Education, 88*(Suppl. 1), S4–S16.

Rennie, L. J., & McClafferty, T. P. (1996). Science centres and science learning. *Studies in Science Education, 27,* 53–98.

Resnick, L. (1987). Learning in school and out. *Educational Researcher, 16*(9), 13–20.

Rickinson, M. (2001). Learners and learning in environmental education: A critical review of evidence. *Environmental Education Research, 7*(Special Issue), 208–318.

Rigano, D. L., & Ritchie, S. M. (2003). Implementing change within a school science department: Progressive and dissonant voices. *Research in Science Education, 33,* 299–317.

Riggs, E. M. (2005). Field-based education and indigenous knowledge: Essential components of geoscience education for Native American communities. *Science Education*, *89*, 296–313.

Roberts, D. A. (1982). Developing the concept of "curriculum emphases" in science education. *Science Education*, *66*, 243–260.

Roberts, D. A. (1983). *Scientific literacy*. Ottawa, Ontario, Canada: Science Council of Canada.

Roberts, D. A. (1988). What counts as science education? In P. J. Fensham (Ed.), *Development and dilemmas in science education* (pp. 27–54). New York: Falmer Press.

Roberts, D. A. (1998, May). *Toward understanding how science teachers think about a new science curriculum policy*. Paper presented at the annual meeting of the Canadian Society for the Study of Education, Ottawa, Ontario, Canada.

Rodriguez, A. J. (2001). From gap gazing to promising cases: Moving toward equity in urban education reform. *Journal of Research in Science Teaching*, *38*, 1115–1129.

Rollnick, M., & Rutherford, M. (1996). The use of mother tongue and English in the learning and expression of science concepts: A classroom-based study. *International Journal of Science Education*, *18*, 91–103.

Roth, W-M., Boutonné, S., McRobbie, C. J., & Lucas, K. B. (1999). One class, many worlds. *International Journal of Science Education*, *21*, 59–75.

Roth, W-M., & Désautels, J. (Eds.). (2002). *Science education as/for sociopolitical action*. New York: Peter Lang.

Roth, W-M., & Désautels, J. (2004). Educating for citizenship: Reappraising the role of science education. *Canadian Journal of Science, Mathematics and Technology Education*, *4*, 149–168.

Roth, W-M., & Lee, S. (2004). Science education as/for participation in the community. *Science Education*, *88*, 263–291.

Roth, W-M., & McGinn, M.K. (1998). <undelete> lives, work, and voices. *Journal of Research in Science Teaching*, *35*, 399–421.

Roth, W-M., Tobin, K., Carambo, C., & Dalland, C. (2004). Coteaching: Creating resources for learning and learning to teach chemistry in urban high schools. *Journal of Research in Science Teaching*, *41*, 882–904.

Rowell, P. M., & Gaskell, P. J. (1987). Tensions and realignments: School physics in British Columbia 1955–1980. In I. Goodson (Ed.), *International perspectives in curriculum history* (pp. 74–106). London: Croom Helm.

Rowland, P. M., & Adkins, C. R. (2003). Native American science education and its implications for multicultural science education. In S. M. Hines (Ed.), *Multicultural science education: Theory, practice, and promise* (pp. 103–120). New York: Peter Lang.

Rubba, P. A. (1987). The current state of research in precollege STS education. *Bulletin of Science, Technology & Society*, *7*, 248–252.

Rubba, P. A. (1989). An investigation of the semantic meaning assigned to concepts affiliated with STS education and of STS instructional practices among a sample of exemplary science teachers. *Journal of Research in Science Teaching*, *26*, 687–702.

Rubba, P. A., & Harkness, W. L. (1993). Examination of pre-service and in-service secondary science teachers' beliefs about science-technology-society interactions. *Science Education, 77*, 407–431.

Rubba, R. A., Schoneweg-Bradford, C., & Harkness, W. F. (1996). A new scoring procedure for the *Views on Science-Technology-Society* instrument. *International Journal of Science Education, 18*, 387–400.

Rubba, P. A., & Wiesenmayer, R. L. (1991). Integrating STS into school science. In S. K. Majumdar, L. M. Rosenfeld, P. A. Rubba, E. W. Miller, & R. F. Schmalz (Eds.), *Science education in the United States: Issues, crises and priorities* (pp. 186–194). Easton, PA: The Pennsylvania Academy of Science.

Rudolph, J. L. (2003). Portraying epistemology: School science in historical context. *Science Education, 87*, 64–79.

Russell, T., McPherson, S., & Martin, A. K. (2001). Coherence and collaboration in teacher education reform. *Canadian Journal of Education, 26*, 37–55.

Russell, T., & Munby, H. (1991). Reframing: The role of experience in developing teachers' professional knowledge. In D. Schön (Ed.), *The reflective turn: Case studies in and on educational practice* (pp. 164–187). New York: Teachers College Press.

Rutherford, F. J. (1988). STS: Here today and . . . ? *Bulletin of Science, Technology & Society, 8*, 125–127.

Ryan, A. G. (1988). Program evaluation within the paradigm: Mapping the territory. *Knowledge: Creation, Diffusion, Utilization, 10*, 25–47.

Ryan, A. G., & Aikenhead, G. S. (1992). Students' preconceptions about the epistemology of science. *Science Education, 76*, 559–580.

Ryder, J. (2001). Identifying science understanding for functional scientific literacy. *Studies in Science Education, 36*, 1–42.

Ryder, J., Hind, A., & Leach, J. (2003a, March). *Enacting lesson resources for teaching about the nature of theoretical models in high school science classrooms.* Paper presented at the annual meeting of the National Association for Research in Science Teaching, Philadelphia, PA.

Ryder, J., Hind, A., & Leach, J. (2003b, August). *Teaching about the epistemology of science in school science classrooms: Case studies of teachers' experiences.* Paper presented at the European Science Education Research Association (ESERA) 2003 conference, Noordwijkerhout, The Netherlands.

Rye, J. A., & Rubba, P. A. (2000). Student understanding of global warming: Implications for STS education beyond 2000. In D. D. Kumar & D. E. Chubin (Eds.), *Science, technology, and society: A sourcebook on research and practice* (pp. 193–230). New York: Kluwer Academic/Plenum Publishers.

Sadler, P. M., & Tai, R. H. (2001). Success in introductory college physics: The role of high school preparation. *Science Education, 85*, 111–136.

Sadler, T. D. (2004). Informal reasoning regarding socioscientific issues: A critical review of research. *Journal of Research in Science Teaching, 41*, 513–536.

Sadler, T. D., Chambers, F. W., & Zeidler, D. L. (2004). Student conceptualisations of the nature of science in response to a socioscientific issue. *International Journal of Science Education, 26*, 387–409.

Sadler, T. D., & Zeidler, D. L. (2004). The morality of socioscientific issues:

Construal and resolution of genetic engineering dilemmas. *Science Education*, *88*, 4–27.

Sáez, M. J., & Carretero, A. J. (2002). The challenge of innovation: The new subject "natural sciences" in Spain. *Journal of Curriculum Studies, 34*, 343–363.

Sáez, M. J., Niño, A. G., Villamañan, R., & Padilla, Y. (1990). Introduction of biotechnology in secondary schools in Spain. In M. Keogh & S. Naidoo (Eds.), *Proceedings of the 9th symposium of IOSTE*: Vol. 1, pp. 558–563. Durban, South Africa: University of Durban-Westville.

Saskatchewan Indian Cultural Centre [SICC]. (1993). *Practising the law of circular interaction*. Saskatoon, Saskatchewan, Canada: Author.

Säther, J. (2003). The concept of ideology in analysis of fundamental questions in science education. *Science & Education, 12*, 237–260.

Scantlebury, K. (1998). An untold story: Gender, constructivism and science education. In W. W. Cobern (Ed.), *Socio-cultural perspectives on science education* (pp. 99–120). Boston: Kluwer Academic.

Schwab, J. J. (1978). *Science, curriculum, and liberal education*. Chicago: University of Chicago Press.

Schwartz, R. S., & Lederman, N. G. (2002). "It's the nature of the beast": The influence of knowledge and intentions on learning and teaching nature of science. *Journal of Research in Science Teaching, 39*, 205–236.

Schwartz, R. S., Lederman, N. G., & Crawford, B. A. (2004). Developing views of nature of science in an authentic context: An explicit approach to bridging the gap between nature of science and scientific inquiry. *Science Education, 88*, 610–645.

Schwarz-Bloom, R. D., & Halpin, J. J. (2003). Integrating pharmacology topics in high school biology and chemistry classes improves performance. *Journal of Research in Science Teaching, 40*, 922–938.

Science Council of Canada [SCC]. (1984). *Science for every student: Educating Canadians for tomorrow's world* (Report No. 36). Ottawa, Canada: Author.

Science Education for Public Understanding Project [SEPUP]. (2003). *SEPUP News*. Berkeley: University of California at Berkeley, Lawrence Hall of Science. Retrieved May 4, 2004, from www.sepup.com.

Seddon, R. (1991). Rethinking teachers and teacher education in science. *Studies in Science Education, 19*, 95–117.

Seiler, G., Tobin, K., & Sokolic, J. (2001). Design, technology, and science: Sites for learning, resistance, and social reproduction in urban schools. *Journal of Research in Science Teaching, 38*, 746–768.

Seker, H. (2004, April). *The effect of using history of science in science teaching*. Paper presented at the annual meeting of the National Association for Research in Science Teaching, Vancouver, British Columbia, Canada.

Semali, L. M., & Kincheloe, J. L. (1999). *What is indigenous knowledge: Voices from the academy*. New York: Falmer Press.

Settlage, J., & Meadows, L. (2002). Standards-based reform and its unintended consequences: Implication for science education within America's urban schools. *Journal of Research in Science Teaching, 39*, 114–127.

Seymour, E. (1995). The loss of women from science, mathematics, and engineering undergraduate majors: An explanatory account. *Science Education*, 79, 437–473.

Shamos, M. (1995). *The myth of scientific literacy*. New Brunswick, NJ: Rutgers University Press.

Shapiro, B. L. (2004). Studying lifeworlds of science learning: A longitudinal study of changing ideas, contests, and personal orientations in science learning. *Canadian Journal of Science, Mathematics and Technology Education*, 4, 127–147.

Shiva, V. (2002). Forward: Cultural diversity and the politics of knowledge. In G. J. Sefa Dei, B. L. Hall, & D. G. Rosenberg (Eds.), *Indigenous knowledges in global contexts: Multiple readings of our world* (pp. vii–x). Toronto: University of Toronto Press.

Shumba, O., & Glass, L. W. (1994). Perceptions of coordinators of college freshman chemistry regarding selected goals and outcomes of high school chemistry. *Journal of Research in Science Teaching*, 31, 381–392.

Shymansky, J. A., Yore, L. D., Treagust, D. F., Thiele, R. B., Harrison, A., Waldrip, B. G., et al. (1997). Examining the construction process: A study of changes in level 10 students' understanding of classical mechanics. *Journal of Research in Science Teaching*, 34, 571–593.

Simonneaux, L. (2001). Role-play or debate to promote students' argumentation and justification on an issue in animal transgenesis. *International Journal of Science Education*, 23, 903–927.

Sjøberg, S. (2000). Interesting all children in "science for all." In R. Millar, J. Leach, & J. Osborne (Eds.), *Improving science education: The contribution of research* (pp. 165–186). Birmingham, UK: Open University Press.

Sjøberg, S. (2003). *Relevance of science education*. Retrieved December 1, 2004, from http://www.ils.uio.no/forskning/rose/.

Smith, L. T. (1999). *Decolonizing methodologies: Research and indigenous peoples*. London: Zed Books.

Smith, M. U., Lederman, N. G., Bell, R. L., McComas, W. F., & Clough, M. P. (1997). How great is the disagreement about the nature of science: A response to Alters. *Journal of Research in Science Teaching*, 34, 1101–1103.

Snively, G. (1990). Traditional Native Indian beliefs, cultural values, and science instruction. *Canadian Journal of Native Education*, 17, 44–59.

Snively, G., & Corsiglia, J. (2001). Discovering indigenous science: Implications for science education. *Science Education*, 85, 6–34.

Solano-Flores, G., & Nelson-Barber, S. (2001). On the cultural validity of science assessments. *International Journal of Science Education*, 38, 553–573.

Solbes, J., & Vilches, A. (1997). STS interactions and the teaching of physics and chemistry. *Science Education*, 81, 377–386.

Solomon, J. (1983). Learning about energy: How pupils think in two domains. *European Journal of Science Education*, 5, 49–59.

Solomon, J. (1984). Prompts, cues and discrimination: The utilization of two separate knowledge systems. *European Journal of Science Education*, 6, 277–284.

Solomon, J. (1988a). Science technology and society courses: Tools for thinking about social issues. *International Journal of Science Education*, 10, 379–387.

Solomon, J. (1988b). The dilemma of science, technology and society education. In P. J. Fensham (Ed.), *Development and dilemmas in science education* (pp. 266–281). New York: Falmer Press.

Solomon, J. (1990). The discussion of social issues in the science classroom. *Studies in Science Education, 18,* 105–126.

Solomon, J. (1992). The classroom discussion of science-based social issues presented on television: Knowledge, attitudes and values. *International Journal of Science Education, 14,* 431–444.

Solomon, J. (1994a). Conflict between mainstream science and STS in science education. In J. Solomon & G. Aikenhead (Eds.), *STS education: International perspectives on reform* (pp. 3–10). New York: Teachers College Press.

Solomon, J. (1994b). Knowledge, values and the public choice of science knowledge. In J. Solomon & G. Aikenhead (Eds.), *STS education: International perspectives on reform* (pp. 99–110). New York: Teachers College Press.

Solomon, J. (1994c). Learning STS and judgments in the classroom: Do boys and girls differ? In J. Solomon & G. Aikenhead (Eds.), *STS education: International perspectives on reform* (pp. 141–154). New York: Teachers College Press.

Solomon, J. (1994d). Toward a map of problems in STS research. In J. Solomon & G. Aikenhead (Eds.), *STS education: International perspectives on reform* (pp. 187–193). New York: Teachers College Press.

Solomon, J. (1994e). Towards a notion of home culture: Science education in the home. *British Educational Research Journal, 20,* 565–577.

Solomon, J. (1997). New science education research for the new Europe? *Studies in Science Education, 29,* 93–124.

Solomon, J. (1998). The science curricula of Europe and notion of scientific culture. In D. A. Roberts & L. Östman (Eds.), *Problems of meaning in science curriculum* (pp. 166–177). New York: Teachers College Press.

Solomon, J. (1999a). Meta-scientific criticisms, curriculum and culture. *Journal of Curriculum Studies, 31,* 1–15.

Solomon, J. (1999b). Science education and the popularisation of science in the new Europe. In M. Bandiera, S. Caravita, E. Torracca, & M. Vicentini (Eds.), *Research in science education in Europe* (pp. 225–233). Boston: Kluwer Academic.

Solomon, J. (2002). The evolution of cultural entities. *Proceedings of the British Academy, 112,* 183–200.

Solomon, J. (2003a). Home-school learning of science: The culture of homes, and pupils' difficult border crossing. *Journal of Research in Science Teaching, 40,* 219–233.

Solomon, J. (2003b). The UK and the movement for science, technology, and society (STS) education. In R. Cross (Ed.), *A vision for science education: Responding to the work of Peter Fensham* (pp. 76–90). New York: RoutledgeFalmer.

Solomon, J., & Aikenhead, G. S. (Eds.). (1994). *STS education: International perspectives on reform.* New York: Teachers College Press.

Solomon, J., Duveen, J., & Scot, L. (1994). Pupils' images of scientific epistemology. *International Journal of Science Education, 16,* 361–373.

Solomon, J., Duveen, J., Scot, L., & McCarthy, S. (1992). Teaching about the nature of science through history: Action research in the classroom. *Journal of Research in Science Teaching, 29*, 409–421.

Solomon, J., Scot, L., & Duveen, J. (1996). Large-scale exploration of pupils' understanding of the nature of science. *Science Education, 80*, 493–508.

Southerland, S. A., Gess-Newsome, J., & Johnston, A. (2003). Portraying science in the classroom: The manifestation of scientists' beliefs in classroom practice. *Journal of Research in Science Teaching, 40*, 669–691.

Spencer, H. (1859). *Education: Intellectual, moral and physical.* New York: John B. Alden.

Spindler, G. (1987). *Education and cultural process: Anthropological approaches* (2nd ed.). Prospect Heights, IL: Waveland Press.

Stairs, A. (1993/94). The cultural negotiation of indigenous education: Between microethnography and model-building. *Peabody Journal of Education, 69*, 154–171.

Stake, R. E., & Easley, J. A. (1978). *Cases studies in science education.* Urbana: University of Illinois, Center for Instructional Research and Curriculum Evaluation.

Stephens, S. (Ed.). (2003). *Culturally responsive science curriculum* (2nd rev.). Fairbanks: Alaska Native Knowledge Network.

Stocklmayer, S. M., Gore, M. M., & Bryant, C. (Eds.). (2001). *Science communication in theory and practice.* Boston: Kluwer Academic.

Stoker, A., & Thompson, P. (1969). Science and ethics: A radical approach to high school science. *Science Education, 53*, 203–209.

Sumida, M., & Kawasaki, K. (2004, July). *A language-culture origin understanding of science in Japan.* Paper presented to the 1st meeting of Culture Studies in Science Education, Kobe University, Japan.

Sutherland, D. L. (2005). Resiliency and collateral learning in science in some students of Cree ancestry. *Science Education, 89*, 595–613.

Sutherland, D. L., & Dennick, R. (2002). Exploring culture, language and the perception of the nature of science. *International Journal of Science Education, 24*, 1–24.

Sutherland, D. L., & Tays, N. (2004, April). *Incorporating indigenous culture into school science.* Paper presented at the annual meeting of the National Association for Research in Science Teaching, Vancouver, British Columbia, Canada.

Sutman, F. X., & Bruce, M. H. (1992). Chemistry in the community: A five year evaluation. *Journal of Chemical Education, 69*, 564–567.

Symington, D., & Tytler, R. (2004). Community leaders' views of the purposes of science in the compulsory years of schooling. *International Journal of Science Education, 26*, 143–148.

Tal, R. T., Dori, Y. J., Keiny, S., & Zoller, U. (2001). Assessing conceptual change of teachers involved in STES education and curriculum development—the STEMS project approach. *International Journal of Science Education, 23*, 247–262.

Taylor, P. C. (2004, July). *Transformative pedagogy for intercultural research.* Paper presented at the 1st meeting of Culture Studies in Science Education, Kobe University, Japan.

Tharp, R. G. (1989). Psychocultural variables and constraints: Effects on teaching and learning in schools. *American Psychologist, 44*(2), 1–11.

Thier, H. D., & Daviss, B. (2001). *Developing inquiry-based science materials: A guide for educators.* New York: Teachers College Press.

Thier, H. D., & Nagle, B. W. (1994). Developing a model for issue-oriented science. In J. Solomon & G. Aikenhead (Eds.), *STS education: International perspectives on reform* (pp. 75–83). New York: Teachers College Press.

Thier, H. D., & Nagle, B. W. (1996). Development and assessment of an issue-oriented middle school science course. In K. Calhoun, R. Panwar, & S. Shrum (Eds.), *Proceedings of the 8th symposium of IOSTE*: Vol. 3, pp. 265–271. Edmonton, Canada: University of Alberta, Faculty of Education.

Thijs, G. D., & van den Berg, E. (1995). Cultural factors in the origin and remediation of alternative conceptions in physics. *Science & Education, 4,* 317–347.

Thomas, G. P. (1999). Student restraints to reform: Conceptual change issues in enhancing students' learning processes. *Research in Science Education, 29,* 89–109.

Thomas, J. (2000). Using current controversies in the classroom: Opportunities and concerns. In R. T. Cross & P. J. Fensham (Eds.), *Science and the citizen for educators and the public* (pp. 133–144). Melbourne: Arena.

Thompson, J. J., Smith, S. S., & Windschitl, M. A. (2004, April). *Seeing beyond science: How underachieving girls engage in personal and relational spaces.* Paper presented at the annual meeting of the National Association for Research in Science Teaching, Vancouver, British Columbia, Canada.

Tobias, S. (1990). *They're not dumb, they're different.* Tucson, AZ: Research Corporation.

Tobin, K. (1997). Cultural perspectives on the teaching and learning of science. In M. Ogawa (Ed.), *Effects of traditional cosmology on science education* (pp. 15–21). Ibaraki, Japan: Ibaraki University, Faculty of Education.

Tobin, K., & McRobbie, C. J. (1996). Cultural myths as constraints to the enacted science curriculum. *Science Education, 80,* 223–241.

Tobin, K., Seiler, G., & Smith, M. W. (1999). Educating science teachers for the sociocultural diversity of urban schools. *Research in Science Education, 29,* 69–88.

Tobin, K., Seiler, G., & Walls, E. (1999). Reproduction of social class in the teaching and learning of science in urban high school. *Research in Science Education, 29,* 171–187.

Traweek, S. (1992). Border crossings: Narrative strategies in science studies and among physicists in Tsukuba science city, Japan. In A. Pickering (Ed.), *Science as practice and culture* (pp. 429–465). Chicago: University of Chicago Press.

Tsai, C-C. (1999). The progression toward constructivist epistemological views of science: A case study of the STS instruction of Taiwanese high school female students. *International Journal of Science Education, 21,* 1201–1222.

Tsai, C-C. (2000). The effects of STS-oriented instruction on female tenth graders' cognitive structure outcomes and the role of student scientific epistemological beliefs. *International Journal of Science Education, 22,* 1099–1115.

Tsai, C-C. (2001a). A science teacher's reflections and knowledge growth about STS instruction after actual implementation. *Science Education, 86,* 23–41.

Tsai, C-C. (2001b). Ideas about earthquakes after experiencing a natural disaster in Taiwan: An analysis of students' worldviews. *International Journal of Science Education, 23,* 1007–1016.

Tytler, R., Duggan, S., & Gott, R. (2001a). Dimensions of evidence, the public understanding of science and science education. *International Journal of Science Education, 23,* 815–832.

Tytler, R., Duggan, S., & Gott, R. (2001b). Public participation in an environmental dispute: Implications for science education. *Public Understanding of Science, 10,* 343–364.

van den Akker, J. (1998). The science curriculum: Between ideals and outcomes. In B. J. Fraser & K. G. Tobin (Eds.), *International handbook of science education* (pp. 421–447). Dordrecht, The Netherlands: Kluwer Academic.

van Driel, J. H., Beijaard, D., & Verloop, N. (2001). Professional development and reform in science education: The role of teachers' practical knowledge. *Journal of Research in Science Teaching, 38,* 137–158.

van Driel, J. H., Bulte, A. M. W., & Verloop, N. (2005). The conceptions of chemistry teachers about teaching and learning in the context of a curriculum innovation. *International Journal of Science Education, 27,* 303–322.

Vázquez-Alonso, A., & Manassero-Mas, M. A. (1999). Response and scoring models for the "Views on Science-Technology-Society" instrument. *International Journal of Science Education, 21,* 231–247.

Venville, G. J., Wallace, J., Rennie, L. J., & Malone, J. A. (2002). Curriculum integration: Eroding the high ground of science as a school subject? *Studies in Science Education, 37,* 43–83.

Wade, P. D., Lederman, N. G., & Bell, R. L. (1997, June). *Assessing understanding of the nature of science: A historical perspective.* Paper presented at the History & Philosophy of Science and Science Teaching Conference, Calgary, Alberta, Canada.

Walberg, H. J. (1991). Improving school science in advanced and developing countries. *Review of Educational Research, 61,* 25–69.

Wallace, J., & Louden, W. (1998). Curriculum change in science: Riding the waves of reform. In B. J. Fraser & K. G. Tobin (Eds.), *International handbook of science education* (pp. 471–485). Dordrecht, The Netherlands: Kluwer Academic.

Wang, H. A., & Schmidt, W. H. (2001). History, philosophy and sociology of science in science education: Results from the third internal mathematics and science study. *Science & Education, 10,* 51–70.

Waring, M. (1979). *Social pressures and curriculum innovations: A study of the Nuffield Foundation science teaching project.* London: Methuen.

Watts, M., Alsop, S., Zylbersztajn, A., & de Silva, S. M. (1997). "Event-centred-learning": An approach to teaching science technology and societal issues in two countries. *International Journal of Science Education, 26,* 341–351.

Weinstein, M. (1998). Playing the paramecium: Science education from the stance of the cultural studies of science. *Educational Policy, 12,* 484–506.

Welch, W. W. (1969). Curriculum evaluation. *Review of Educational Research, 39,* 429–443.

Welch, W. W. (1973). Review of the research and evaluation program of Harvard Project Physics. *Journal of Research in Science Teaching, 10,* 365–378.

Welch, W. W. (1979). Twenty years of science curriculum development: A look back. *Review of Research in Education, 7*, 282–306.

Welch, W. W. (1995). Student assessment and curriculum evaluation. In B. J. Fraser & H. J. Walberg (Eds.), *Improving science education* (pp. 90–116). Chicago: The National Society for the Study of Education (University of Chicago Press).

Welch, W. W., & Rothman, A. I. (1968). The success of recruited students in a new physics course. *Science Education, 52*, 270–273.

Welch, W. W., & Walberg, H. J. (1967). Are the attitudes of teachers related to declining percentages of enrollments in physics? *Science Education, 51*, 422–436.

Welch, W. W., & Walberg, H. J. (1972). A national experiment in curriculum evaluation. *American Educational Research Journal, 38*, 373–383.

Welzel, M., & Roth, W-M. (1998). Do interviews really assess students' knowledge? *International Journal of Science Education, 20*, 25–44.

Wertsch, J. V. (1991). *Voices of the mind: A sociocultural approach to mediated action.* Cambridge, MA: Harvard University Press.

White, R. (2001). The revolution in research on science teaching. In V. Richardson (Ed.), *Third handbook of research on teaching* (4th ed., pp. 457–470). Washington, DC: American Educational Research Association.

Wiesenmayer, R. L., & Rubba, P. A. (1999). The effects of STS issue investigation and action instruction versus traditional life science instruction on seventh grade students' citizenship behaviors. *Journal of Science Education and Technology, 8*, 137–144.

Wilson, M. (2004). *Berkeley evaluation & assessment research center.* University of California at Berkeley. Retrieved December 1, 2004, from http://bear.berkeley.edu/pub.html.

Wilson, S. (2003). Progressing toward an Indigenous research paradigm in Canada and Australia. *Canadian Journal of Native Education, 27*, 161–178.

Winther, A. A., & Volk, T. L. (1994). Comparing achievement of inner-city high school students in traditional versus STS-based chemistry courses. *Journal of Chemical Education, 71*, 501–505.

Wolcott, H. F. (1991). Propriospect and the acquisition of culture. *Anthropology and Education Quarterly, 22*, 251–273.

Wynne, B. (1991). Knowledge in context. *Science, Technology & Human Values, 16*, 111–121.

Yager, R. E. (Ed.). (1996). *Science/technology/society as reform in science education.* Albany: State University of New York Press.

Yager, R. E., & Krajcik, J. (1989). Success of students in a college physics course with and without experiencing a high school course. *Journal of Research in Science Teaching, 26*, 599–608.

Yager, R. E., Snider, B., & Krajcik, J. (1988). Relative success in college chemistry for students who experienced a high school course in chemistry and those who had not. *Journal of Research in Science Teaching, 25*, 387–396.

Yager, R. E., & Tamir, P. (1993). STS approach: Reasons, intentions, accomplishments, and outcomes. *Science Education, 77*, 637–658.

Yeo, R. (1981). Scientific method and the image of science, 1831–1890. In

R. MacLeod & P. Collins (Eds.), *The parliament of science* (pp. 65–88). North-wood, England: Science Reviews.

Yerrick, R., Parke, H., & Nugent, J. (1997). Struggling to promote deeply rooted change: The "filtering effect" of teachers' beliefs on understanding transformational views of teaching science. *Science Education, 81,* 137–159.

Zacharia, Z., & Calabrese Barton, A. (2004). Urban middle-school students' attitudes toward a defined science. *Science Education, 88,* 197–222.

Zeidler, D. L. (1997). The central role of fallacious thinking in science education. *Science Education, 81,* 483–496.

Zeidler, D. L. (Ed.). (2003). *The role of moral reasoning on socioscientific issues and discourse in science education.* Boston: Kluwer Academic.

Zeidler, D. L., Walker, K. A., Ackett, M. A., & Simmons, M. L. (2002). Tangled up in views: Beliefs in the nature of science and responses to socioscientific dilemmas. *Science Education, 86,* 343–367.

Ziman, J. (1980). *Teaching and learning about science and society.* Cambridge: Cambridge University Press.

Ziman, J. (1984). *An introduction to science studies: The philosophical and social aspects of science and technology.* Cambridge: Cambridge University Press.

Zohar, A., & Nemet, F. (2001). Fostering students' knowledge and argumentation skills through dilemmas in human genetics. *Journal of Research in Science Teaching, 39,* 35–62.

Index

NAMES

SUBJECTS

About the Author

Glen Aikenhead developed a keen interest in a humanistic perspective on science when he was an undergraduate honors chemistry student at the University of Calgary, Canada, and later as a research chemist. His master's and doctoral studies in science education at Harvard University enhanced this humanistic perspective, which he put into practice as a science teacher at international schools in Germany and Switzerland. At the University of Saskatchewan, Saskatoon, Canada, where he is currently a Professor of Education, this humanistic perspective evolved as he conducted research into curriculum policy, student assessment, classroom materials, classroom instruction, and more recently, cross-cultural science education. Glen's collaboration with high school teachers, students, and fellow scholars produced such as publications *Science: A Way of Knowing* (1975), *Science in Social Issues: Implications for Teaching* (1980), *Views on Science Technology and Society* (VOSTS, 1989), *Logical Reasoning in Science & Technology* (1991), *STS Education: International Perspectives on Reform* (1994), and *Rekindling Traditions: Cross-Cultural Science & Technology Units* (2000). He is an active member in the International Organization for Science and Technology Education, the National Association for Research in Science Teaching, the American Educational Research Association, and the European Science Education Research Association.